Amateur Radio
HF ANTENNAS

VE2DPE's
Four-Book Collection

by Claude Jollet
First edition published November 2016
Copyright © 2016 Claude Jollet

BOOK ISBN: 978-0-9950273-5-0
MOBI ISBN: 978-0-9950273-3-6
EPUB ISBN: 978-0-9950273-4-3
PDF   ISBN: 978-0-9950273-2-9

CIP Data on file with the National Library of Canada.

Legal deposit: fourth quarter 2016
Quebec National Library And Archives

About This e-Book

The contents of this book are mostly aimed at the amateur radio beginner and aspiring ones. Therefore, this book provides answers to basic questions like: What is the best HF antenna for my needs and location? What type of stand-alone antenna tuner should I use and which should I avoid? How can I hide my HF antenna from the neighbors and still get acceptable performance from it? What about lightning protection?

This book will supply immediately useful answers to the above questions and many more.

A properly designed and installed amateur radio HF antenna system can potentially make the humblest ham radio equipment perform like stations worth thousands of dollars.

We are confident that the antenna experimenter will find the information given here priceless. Furthermore, any ham radio operator, armed with the information this book contains, will become a much better informed buyer of commercially made HF antenna systems and accessories.

This special edition is published in response to ham radio

operators who wrote to ask that all the basic information, on and related to amateur radio HF antennas, be made available in one book instead of four, arguing that it would be more convenient. The author and publisher agree. Therefore this edition contains the complete four-book series on Amateur Radio HF Antennas published by Claude Jollet, VE2DPE.

About VE2DPE

Claude Jollet obtained his amateur radio *Basic Certificate*
with his operator's license in 1973 and his *Advanced
Certificate* in 1974. He has been operating under the call
sign VE2DPE from day one. He loves everything about
ham radio, but he especially loves to experiment with
antennas. Hence his decision to share the fruits of his more
than 40 years experience in this e-book and the others in
the series. As the call sign reveals, Claude lives in the
predominantly French speaking Canadian Province of
Québec, in a small town called Notre-Dame-des-Prairies.

Dear Reader

*** BOOK ONE ***

HF Antenna Basics

About e-Book One

What stands between you, the amateur radio operator, and all the other amateur radio operators in the world? Antennas. Yours, theirs…and propagation conditions between. In other words, you can have the most advanced transceiver, coupled with the most powerful amplifier on the market, if you know too little about how antenna systems work and about radio wave propagation, you risk spending your time and money for naught. This e-book is the first of a series on HF antennas. Together, they will help ensure that you turn the situation around to your advantage.

INTRODUCTION
Ham Radio HF Antennas

All ham radio antennas involve compromises. There are no exceptions. This e-book will show you how to choose the set of compromises that will best fit your particular situation and requirements.

Most ham radio operators use the same antenna for receiving and transmitting on a given amateur radio band. That is the first compromise of many!

The greater the number of frequency bands you want to work with the same antenna system, the greater the number of compromises you will have to live with. The high-performance yagi type antenna in the picture (below) is one of the best set of compromises available for a multi-band operation on HF.

But few of us have the money for a multi-element directional antenna (a yagi or quad) covering all HF bands from 20 to 10 meters, not to mention the tower to support it. Fewer still have the space or the money to have individual antennas for each band!

A Setup To Avoid!

Here is a typical example of the worst possible setup, all too often encountered on the HF bands:

You should avoid being the amateur radio operator

calling "CQ" while feeding maximum legal power into a multi-band trap dipole or, worse, a multi-band trap vertical!

If you **are** such an operator, you will often not "hear" the hams answering your calls! Why? Because of the poor receiving efficiency of such multi-band antennas. Even when installed properly, they may "appear" to be "effective" radiators, but be advised that they are very inefficient. Under full legal transmitting power, the signal from such "inefficient" antennas can be detected so far away that the same antenna cannot detect the signal of the DX (far distant) station responding to the call!

What is the point of operating with an antenna system that does not let you hear those who can hear you and are answering your call?

The Ideal Setup

You can easily avoid unbalanced operating conditions such as I described above.

- Reduce transmitting power to correspond to the receiving capabilities of the amateur radio antenna you are using. Remember that a full length half-wave dipole will capture much more RF energy (on receive) than a shortened trap dipole on the same frequency band. Traps introduce resistance and reactance. Part of the RF energy is transformed into heat by the traps. Whatever is left will flow down the transmission line to your receiver or transceiver.

- If you can, use a separate, high performance receiving antenna on each HF band you want to work.

- Or, use a beam-type directional antenna, on your ham radio tower, for both transmitting and receiving (as in the picture shown previously).

If you like to experiment, you can even design your own "dream antenna" that will fit your operating conditions and preferences. You will find all you need to know in the ARRL Antenna Book, including how to use an antenna design program to build your next high-performance homemade ham radio antenna.

CHAPTER ONE

Antenna Selection Tips

Every ham radio antenna is full of inevitable compromises. Some antennas have more than others as we will see here. When choosing or building an amateur radio antenna, the most common compromises you have to make will fall in the following categories:

- Cost (for a commercially made antenna or cost of parts if homemade).
- Available space (both horizontal and vertical).
- Durability.
- Performance (of course!).
- City bylaws (increasingly…sigh!).

The above are by no means the only compromises one must make, but addressing them, and finding the best solution for my own needs constitute the kind of challenge I enjoy.

In Search Of The "Perfect" Antenna

If perfect antennas were possible I would make them and become a millionaire! The bad news is that the perfect antenna does not exist, even in theory! The theoretically perfect antenna can never be built...because theory itself is not perfect! However, the good news is that experimenting with homemade antennas is one of the most accessible and enjoyable aspects of amateur radio.

Homemade Ham Radio Antenna

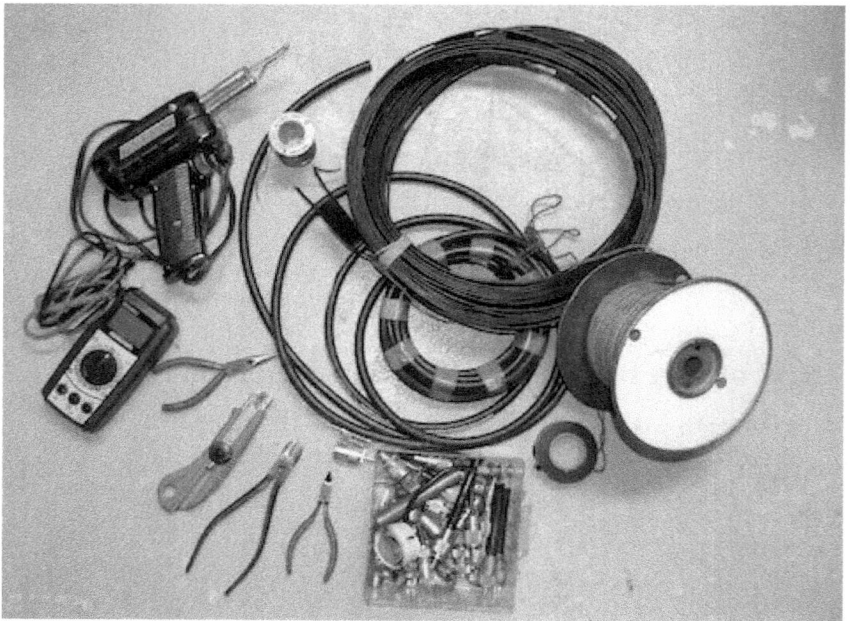

Nothing is more satisfying than building a few antenna prototypes and getting better results with each new one. Trust me. I know because experimenting with homemade antennas is the part of the ham radio hobby that I love the most. I have been tinkering with antennas since 1973.

The aspects I love best are:

1. Studying (learning) antenna theory.
2. Researching and studying experiments made by other hams.
3. Designing my own ham radio wire antenna with its own optimized set of compromises.
4. Building, erecting and testing it on the air.
5. Then starting all over on a new antenna project!

I especially love experimenting with wire antennas of various configurations, including so-called "stealth" (relatively invisible) antennas. In fact, when you hear me on the HF bands, it is likely because I am testing the performance of my latest "baby". I will give you an example, in my soon to be released **Antenna Book Three**, of one of my experimental antennas, the *"VE2DPE 160M Special"*.

What if you do not have the necessary horizontal space to put up a full size 160 meter ham radio dipole? Then try your hand at building a much smaller...

- ham radio dipole for 10 meters
- or a folded dipole for 6 meters.

These are easy and fun to build and use on the air!

If you do not have the real estate to put up a "classic" half-wave horizontal dipole, do not despair! I will give you some space-saving configurations of ham radio HF antennas that might do the trick in **Book Two**.

CHAPTER TWO

The Classic Dipole

The ham radio dipole is called a half-wave antenna because its length corresponds to an electrical half wave at the frequency for which it is intended. The dipole will always be physically shorter than its "free space" length. Why? First, the resistance in the antenna wire slightly impedes the flow of RF energy. Second, impedance and reactance also slowdown the flow of RF current. These are introduced by:

- ground conductivity below the antenna,
- its height above ground,
- and the presence of buildings and trees in the vicinity.

The dipole is called a balanced antenna because it is "fed" at its exact center. In other words, the transmission line is connected at the center. The majority of amateur radio operators throughout the world choose this antenna over other types, at least, to start.

Why Is The Dipole So Popular?

Because it is an inexpensive, yet effective, antenna. Hams can cut costs further by building it themselves using wire, insulators and rope that they already have in their (or a friend's) "junk box" of spare parts!

A Picture Is Worth...

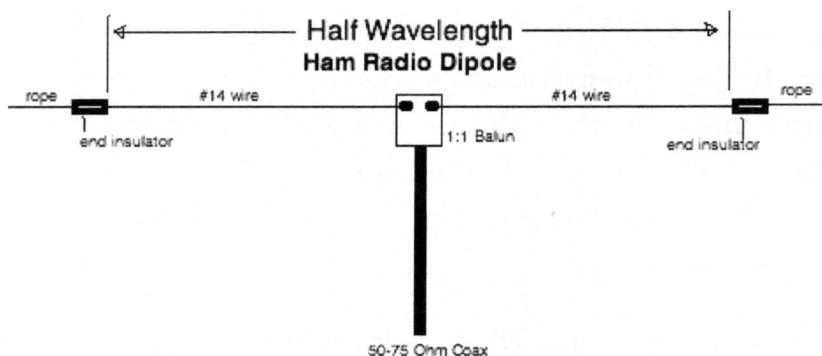

In "free space", the characteristic impedance of a half wave dipole is around 73 ohms and it's length, in meters, is obtained by this formula:

length in meters = (299.7925 / frequency in MHz) / 2

Thus a dipole on the 20 meter band would measure (in free

space):
(299.7925 / 14) / 2 = 10.706875 meters or 35.127543 feet
long.

However, if the horizontal dipole is between 0.1 and 0.2
wavelength above ground, its impedance will be somewhat
lower and closer to 50 ohms, which is the characteristic
impedance of commonly used coaxial transmission line
such as RG-8, RG-8X and RG-58.

As a bonus, the impedance will remain low at odd
harmonic multiples of the basic frequency! Yes, you can
operate a 40 meter (7 MHz) dipole on 15 meters (21 MHz)
too. A two-for-one antenna!

The Antenna "System"

Bear in mind that the antenna is only part of a coherent
system! An antenna system is composed of:

- The antenna (the radiating and receiving portion).
- The transmission line (carrying the RF energy
  between the antenna and the transceiver).
- The impedance matching unit between the
  transceiver and the transmission line (if one is
  used).

The transceiver or receiver will "see" the above three components as a whole.

Dipole Formula

The formula to calculate the (approximate) overall **physical length** of a dipole is:

Length (meters) = 142.58 / frequency (MHz)

Length (feet) = 468 / frequency (MHz)

Thus, the operating length of a half wave dipole on the 20 meter band (14 MHz) would be:

468 / 14 = 33.4286 feet (1.7 feet *shorter* than in free space)

The length of wire obtained by the formula is cut in half. The transmission line connects between each half of wire, which becomes the dipole's feed point.

It is worth repeating here that the above formula gives an approximate result because the length at the desired resonant frequency is affected by its operating environment conditions, such as:

- the antenna height above ground,
- the conductivity of the ground itself,

- the proximity of metal structures and other objects such as trees and power lines.

Dipole *Resonant* Length

Most dipoles will require a little "pruning" (on site) to resonate at the desired frequency. I recommend that you:

1. Cut your dipole wire some 2-3% longer than the length given by the formula.
2. Make a note of the length obtained in step 1.
3. Raise the dipole to its operating height.
4. Measure the SWR at several frequencies within the intended frequency band. (Use only a few watts and pick a quiet time on the band to make your tests).
5. Note the frequency (F_min) at which minimum SWR is obtained.
6. Multiply (F_min) by the antenna length recorded in step 2.
7. Divide the result of the above multiplication by the desired frequency of operation, to obtain the final length.
8. Trim both ends of the dipole down to the final length obtained in step 7.

Transmission Line

You can feed your dipole with coax, as already mentioned above. However, coaxial cable is unbalanced! When feeding a balanced load, such as a dipole, with an unbalanced transmission line, the antenna will induce RF currents on the outer shield of the coax during transmission.

These unwanted RF currents spell nothing but trouble, not the least of which are RFI (radio frequency interference) caused by stray RF energy being radiated back in the shack to wreak havoc with your computer or other delicate electronic devices!

Fortunately, you can prevent these unwanted RF currents from traveling back on the outer shield of your coaxial transmission line with a RF choke.

RF Choke Balun

To make a choke balun, all you need to do is wind a portion of the coaxial transmission line to form a coil. A choke balun made of coax is most effective when a single layer is close-wound on a form, such as 4 inch plastic drainpipe or 6 inch "schedule 40" PVC pipe. The tables below list values for each HF amateur band. Form size and number of turns are optimized for each band.

Single Band RF Chokes
(most effective)

Band (meters)	Form (inches)	Coax RG-213 RG-8	Coax RG-8X RG-58
160	6 in.	8 turns	5-6 turns
80	6 in.	8 turns	5-6 turns
40	6 in.	8 turns	5-6 turns
30	6 in.	8 turns	5-6 turns
20	4 in.	12 turns	7-8 turns
17	4 in.	12 turns	7-8 turns
15	4 in.	6 turns	4-5 turns
12	4 in.	6 turns	4-5 turns
10	4 in.	6 turns	4 turns

Multi-Band RF Chokes

Freq. Range (MHz)	COAX: RG-8, RG-58, RG-59, RG-8X, RG-213
3.5 - 30	3.05 m. (10 ft.), 7 turns
3.5 - 10	5.49 m. (18 ft.), 9-10 turns
14 - 30	2.44 m. (8 ft.), 6-7 turns

Alternatively, you could use Amidon(TM) *31 material* cable clamp ons. These molded ferrite clamp ons come in different sizes and shapes to fit over RG-58, RG-59, RG-8 coax cables. The *31 material* ferrite formula is designed to stop RF, in the range between 1 to 300 MHz, from travelling back in the shack along the outer shield of the coaxial transmission line. I install two to four of these (as needed) on the coax, outdoors, just before the cable enters the outside wall to my radio room.

Dipole Hardware

Many types of wire, insulators and rope can be pressed into service for a temporary installation. But for a permanent and safe all-weather installation, here is what I recommend.

Wire

I use either 14 gauge stranded (7x22) hard-drawn copper wire, for spans less than 45 meters (150 ft.) between **stable** supports. But when a dipole antenna needs to be strung between trees, I recommend the very strong VariFlex(TM) 13 gauge, 19 strand, copper-clad steel wire. I use the latter for 80 meter dipoles that I install between the maple trees that I'm fortunate to have on either side of my house.

Insulators

You can use a ceramic *"dog bone"* as center insulator, but you will have to wrap your coax around it, tie it securely with tie-wraps, then split the center conductor and braid to connect to each side of your ham radio dipole, then seal to prevent water from seeping in the coax! Not exactly the best setup.

I prefer using a center insulator with a SO-239 which makes it much easier to seal against water infiltration. I like to use Coax-Seal(TM) for this purpose.

Of course, you could use a commercially made 1:1 balun at the dipole feed point. It will serve as center insulator and coax connector. The balun will improve the radiation pattern somewhat ... if your dipole is at least 1/4 wavelength above ground.

Finally, I use Delta CIN ceramic end insulators by Alpha Delta Communications, at each end of a dipole, for their resistance to RF and long leakage path. Remember that high voltage is present at each end of a dipole while transmitting!

Rope

I have had a dipole strung between two large trees for years with (3/16 in.) Mil Spec Dacron® rope. My antenna is still up there! This rope is very strong, abrasion and UV resistant.

A Reminder

Take your ham radio dipole down at least once a year to check for damage (frayed ropes, damaged coaxial shield or connection, etc). Repeat this inspection more than once a year if you have had severe weather.

— 73 —

*** BOOK TWO ***

HF Antennas
For
Limited Space

About e-Book Two

What additional challenge is there to HF communications besides:

- Overcoming often adverse propagation conditions;
- Establishing solid contacts in the cacophony of pileups when conditions are good;
- Acquiring the necessary operating skills?

Finding space to install a large HF antenna, that's what!

This e-book, with its many illustrations, will supply some of the most viable HF antenna solutions for restricted spaces, and explain the pros and cons of each. You will then be free to choose the solution that will "fit" into the space you have available while meeting most of your needs.

The solution you opt for will involve compromises. But, at least, you will have the luxury of being able to choose.

CHAPTER ONE

"Full-Size" Solutions

The amateur radio HF antenna, of the half-wave dipole variety, requires a fair amount of horizontal space, especially on the lower amateur bands of 160M, 80M and 40M.

It is possible, however, to "fit" a full-length half-wave dipole in less horizontal space than they would normally require, while retaining most, if not all the advantages of the "classic dipole" described in **Book One** of this series.

Below are a few possible configurations. One of them even has a slight performance edge for DX (long-distance contacts) over the horizontal dipole!

The half-wave HF antennas described below do not impose the compromises of inductively "loaded" antennas (using loading coils), namely: lower efficiency and narrower useful bandwidth.

WARNING!

1. High voltage is present at the ends of half-wave dipole antennas when transmitting.

2. The ends of the antennas, described in this chapter, are lower above ground than normal!

3. Take all necessary precautions to prevent anyone from coming in contact with your antenna!

The Half-Wave "Full Sloper" Dipole

This amateur radio HF antenna needs only one tall support and takes less horizontal space than one installed horizontally.

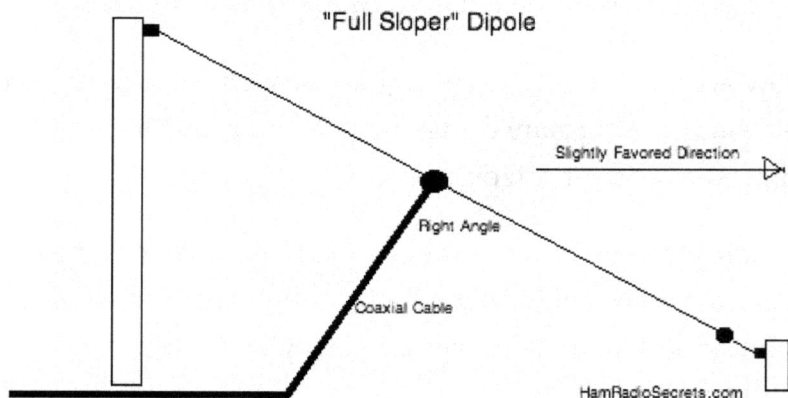

"Full Sloper" Dipole

Slightly Favored Direction

Right Angle

Coaxial Cable

HamRadioSecrets.com

One end of the dipole is tied to the top of a tree (or other tall support). The other end slopes down at a convenient angle to a second support, lower in height but remaining at least 7-8 feet above ground, to prevent anyone from accidentally touching the lower end of the sloper while it is transmitting.

The radiation pattern will be almost omnidirectional, with a slight advantage toward the "sloping" direction.

The coaxial transmission line should be positioned at a 90 degree angle to the "sloper" for at least 1/4 wavelength,

- to avoid distortion in the radiation pattern,
- and to avoid RF energy from traveling back to the shack over the coax shield! (You can eliminate these unwanted RF currents with common-mode RF chokes, as described in **Book One** of this series).

NOTE: You don't have enough space for a full half-wave sloper on 160 or 80 meters? Don't give up! The inverted "V" or "U" configurations offer an excellent solution for restricted spaces.

The Inverted "V" Dipole

The inverted "V" dipole configuration is also a space saver! For best performance, the angle between the two legs should be greater than 90 degrees.

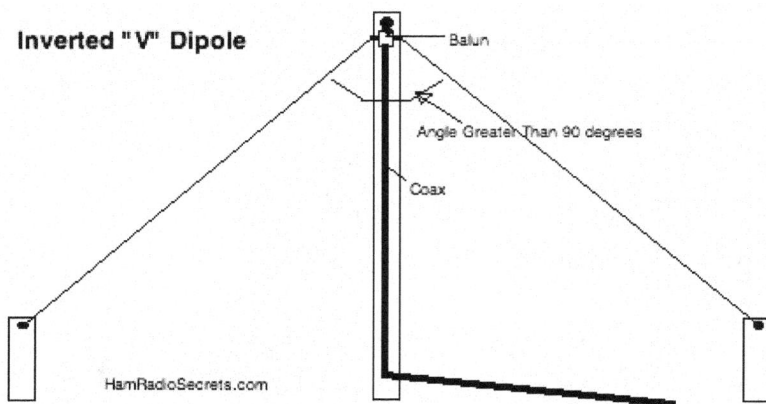

Inverted "V" Dipole

Balun

Angle Greater Than 90 degrees

Coax

HamRadioSecrets.com

If you feed this antenna directly with a coaxial cable, without a balun at the antenna feed point, you might experience RF feedback in the shack. In that case, you will need to insert one, perhaps two, common-mode RF chokes in the transmission line. These are needed to "choke off" unwanted RF currents, induced by the radiating legs of the antenna onto the transmission coaxial shield. (See Book One for information on RF chokes for HF).

The Inverted "U" Dipole

Most of the RF energy radiated and captured by a half-wave dipole antenna is concentrated in the middle 60% portion of it.

Inverted "U" Dipole

Rope — Antenna — Balun — Coax — Antenna — Rope

Antenna — Antenna — Antenna — Antenna

Rope — Fixed Anchoring Point — Fixed Anchoring Point — Rope

HamRadioSecrets.com

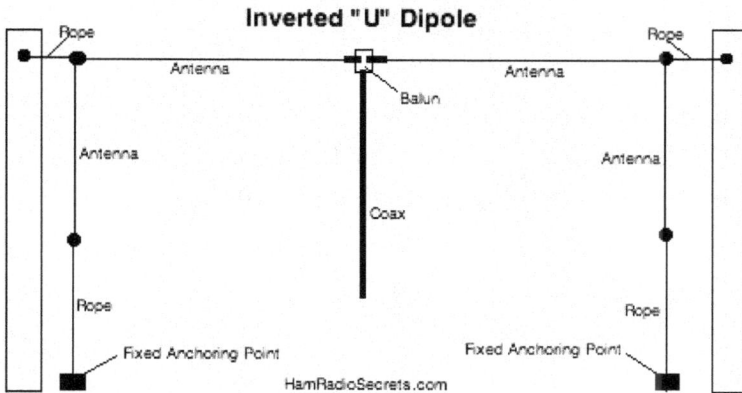

The ends of the dipole can be "dropped" down from the horizontal, beyond the central 60% of the antenna, without much adverse effect.

Therefore, the dipole could be strung between two tall supports that are less than half a wavelength apart at the lowest operating frequency. (NOTE: A dipole antenna can be operated on its odd multiple harmonics. In other words, a 7 MHz dipole can also operate on 21 MHz).

The horizontal portion of this amateur radio HF antenna should be at least 60% of the overall length, for best performance.

"Low-Profile" HF Antennas

In **Book Three** of this series, I will give you techniques to install "low-profile", sometimes called "stealth", outdoor antennas for HF when city bylaws (or uncooperative neighbors) will not otherwise allow one. Book Three will also cover how to install an antenna for HF indoors, out of sight. The book will include candid coverage of the dangers and pitfalls of an indoor antenna, as well as the precautions you must take to minimize them.

CHAPTER TWO

Short HF Antennas

You may have heard that short amateur radio antennas are not as efficient, nor as effective as a "full length" half-wave dipole, regardless of the configuration chosen.

Good news! I describe in this article the only exception, which I know, that actually works!

To be honest, even if effective, the antenna described below will tune somewhat more sharply (have narrower 2:1 SWR bandwidth) than a full half wave dipole.

When you don't have the space, you have to accept a few compromises. Bandwidth is one of them.

Nonetheless, this "shortened" antenna will not disappoint you! John Jensen, OZ3PAX, built one for the 17 meter band. Here is his comment:

"I am quite pleased with my Linear Loading Short Ham Antenna ... I built it from 450 ohm ladderline and it works great. I built it

for the 17 meter band, but it also tunes nicely on the 15 and 20 meter band."

I believe the details of this technique were first published some years ago by John Stanford, NN0F.

Linear-Loaded Short HF Dipole

For many amateur radio operators, it is not feasible to put up a full-length dipole on HF.

But linear-loading a dipole just might fit into your available space! A linear-loaded dipole, as illustrated below,

- is about 30-35% shorter than a "classic half-wave dipole" at the same frequency of resonance;
- has a radiation resistance around 35 Ohms (will require an impedance matching tuner);
- is just as effective as a "full-length" half-wave dipole!

"Linear-Loaded" Short HF Dipole

390 Ohm open wire "ladder line"   390 Ohm open wire "ladder line"

1:1 Balun

Coax

HamRadioSecrets.com

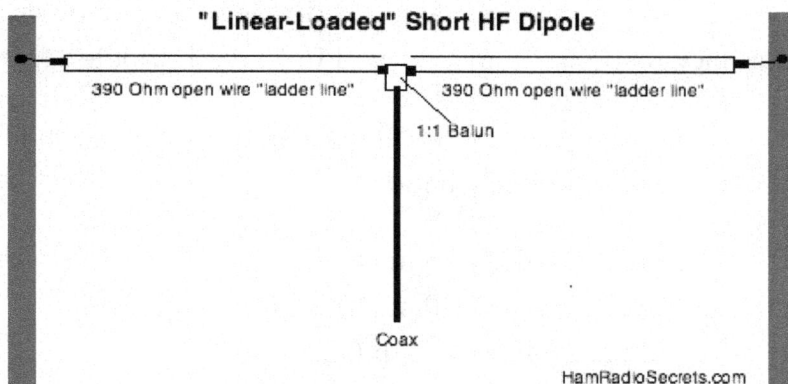

Note that it "looks" like a folded dipole, but the top part is open! You should install a ceramic end-insulator in the opening for added mechanical strength. (The insulator is omitted in the drawing to make the opening obvious).

Linear-Loaded Dipole Dimensions

Here are approximate dimensions for a linearly loaded dipole for each amateur radio band. These dimensions are intentionally slightly long! You will have to obtain the final dimensions experimentally, on-site, by "pruning" to resonance.

Have you forgotten the technique to prune a dipole to resonance? You will find the details in **Book One** of this series on HF antennas. But, be careful to prune in *smaller* *increments* because a linear-loaded dipole has a narrower frequency response than a normal full-length dipole.

Here are the dimensions you should start with for each HF amateur radio frequency band:

- 10 M (28.5 MHz) 3.5 m (11.5 ft)
- 12 M (24.9 MHz) 4.0 m (13.2 ft)
- 15 M (21.1 MHz) 4.73 m (15.5 ft)
- 17 M (18.1 MHz) 5.51 m (18.1 ft)
- 20 M (14.1 MHz) 7.08 m (23.2 ft)
- 30 M (10.1 MHz) 9.89 m (32.44 ft)
- 40 M (7.1 MHz) 14.06 m (46.14 ft)
- 80 M (3.6 MHz) 27.74 m (91.0 ft)
- 160 M (1.85 MHz)53.97 m (177.08 ft)

Construction Of Linear Loaded Antennas

I use commonly available 390 Ohm "ladder line" with #14 stranded, copper-clad conductors. It is sturdy, and lasts for years. 450 Ohm ladder line will work just as well.

For the central "attachment" I use two LadderLoc(R) insulators, head to head, to form a center insulator and another to support the ladder transmission line. The three LadderLoc(R) insulators form a "T" at the center of the dipole.

I recommend 3/16 in. Mil Spec Dacron(R) rope to tie the ends to tall supports such as trees. This rope is very strong and UV resistant.

Bonus Configurations

Linear-loaded antennas do not have to be limited to horizontal installations! You can save even more space by installing them as "sloper", inverted "V" or inverted "U" configurations, as described earlier in this book.

CHAPTER THREE

The Dipole on "Coils"

The use of coils (inductors) is a very popular method of making a half-wave dipole resonate at a given frequency, while being physically shorter than normal. This type of antenna is suitable for single band, narrow bandwidth use.

The Technique

By replacing part of the normal dipole length with inductance, in the form of loading coils, we can make a shorter half-wave dipole become the "electrical" equivalent of a full-size half wavelength dipole.

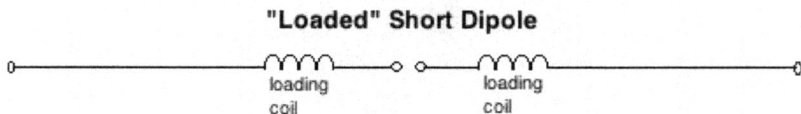

"Loaded" Short Dipole

loading coil   loading coil

www.HamRadioSecrets.com

However, the resulting "loaded" short dipole becomes a set of compromises. Fortunately, careful design can make these compromises "acceptable". What are the compromises or trade-offs?

- When using a "loaded" dipole, you are trading some performance for the capability of being able to install a dipole in less space than normal.

- The loading coils introduce resistive and reactive components. These will cause a portion of the RF power fed to, or received by, the "loaded" antenna, to be lost in the form of heat, instead of being radiated, or captured, as useful signal!

- The resulting loss of a portion of the RF energy may not be critical when in transmit mode but, in some cases, it can be quite significant when trying to receive weak signals!

Factors To Consider

Here are some of the factors one must take into consideration when designing and building, or buying, a "loaded" short dipole.

- The positioning of the loading coils along the wire

is critical. Ideally, greatest efficiency is obtained with coils near the ends of the dipole. But then coils would have to be impossibly large (infinite size!) in terms of inductance! The loading coils of a short dipole must therefore be positioned away from the ends, but still be closer to the ends than to the center feed point.

- Close winding of coils should be avoided!

- The wire size (gauge) used for the coil is a factor. The smaller the wire gauge, the more resistance is introduced, because resistance dissipates RF energy as heat, not as signal! Coils using small diameter copper tubing would be best, but would introduce mechanical problems in the construction of the antenna.

- The quality of the wire is another factor. Solid copper is ideal. Copper-clad steel wire is less desirable.

- Air-wound coils would be best but are often not mechanically feasible in this type of application. So coils must be wound on a form. The insulating properties of the form material are a factor, at radio frequencies in the HF range. Coils wound on glass or porcelain heat up much less than coils wound on plastic material. A loaded antenna will radiate more

RF energy, and capture more RF on receive, when loading coils are of high quality.

- Another consideration is weatherproofing. It is relatively easy to weatherproof air-wound coils against rain. But when the coils are wound on forms, they must be weatherproofed because water will short-circuit the coils. Also, think of sticky snow or even freezing rain! A "loaded" dipole will resonate off frequency, and its feed point impedance will be different, when used with a coating of ice or snow on the loading coils.

- If you feed enough power to the antenna, coils will heat up and eventually melt the ice. But be careful. Even an automatic antenna tuner will have trouble keeping up with a good match as the frequency of resonance and impedance drift up and down while the antenna is melting the ice!

- Finally, coils are the "high-Q" components of the loaded dipole. They are designed for a given frequency. This means that when you are using a "loaded" dipole away from its resonant frequency, the antenna's efficiency and effectiveness drop dramatically!

- The shorter the antenna (physically) with respect to its intended wavelength (the larger the inductance of the coils) the worse things get! Loaded dipoles are very narrow-band antennas, best suited for use on a narrow portion of a frequency band. They are also less effective on receive. A loaded dipole, being smaller than a normal full-size dipole, will capture less of the incoming radio wave energy on its intended wavelength of operation.

The ideal (most efficient) coil is:

- air wound (not wound or supported on a form),
- as large as it is long (length to diameter ratio of one),
- made of solid copper wire. Minimum wire size (gauge) must be chosen for the maximum current that is expected to flow through it.

The *"ideal"* antenna radiates 100% of the RF energy fed into it, and transfers to the transmission line all the RF energy that strikes it. But the *"ideal"* antenna does not exist!

Fortunately, a few manufacturers do make a very acceptable "loaded" short dipole. Their design and craftsmanship reduce the trade-offs to the most acceptable levels possible, while remaining relatively effective and affordable!

The Short Dipole
For Multiband Use

Some dipoles are designed to be used on more than one band of frequencies. Multiband dipoles often use RF "traps". Do not confuse loading coils and frequency traps. They may look alike, when encased in a weatherproof sleeve, but they do not have the same electrical characteristics.

Discussion of traps is beyond the scope of this e-book. But I will at least add this: an RF "trap" consists of a coil and a capacitor connected in parallel to form a tuned circuit. Values for the inductor and capacitor are carefully chosen to resonate at a given frequency. The traps are inserted in series, at very precise locations, on each side of a dipole (or in the portion above ground of a vertical "monopole" antenna).

For example, let's consider a dipole designed for the 80 meter band (3.5 MHz) and the 40 meter band (7.0 MHz). The antenna will use traps that are resonant at 7.0 MHz.

When the dipole is used on 40 meters, the traps offer a high impedance at 40 meters and act as insulators, rendering inactive the portions of the antenna lying beyond the traps. When the antenna is used on the 80 meter band, the traps offer no resistance, and the overall length of the antenna becomes active. In fact, the traps become inductive at 80 meters and thus contribute to the electrical length, making the antenna (slightly) physically shorter than a normal dipole on 3.5 MHz.

80 meter (3.5 MHz) Trap Dipole

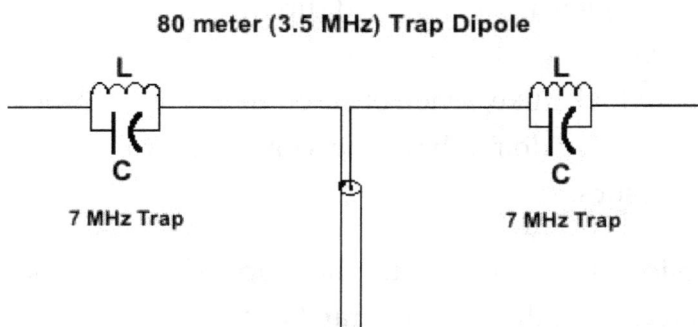

7 MHz Trap   7 MHz Trap

Let me point out that "traps" are even more inefficient ("lossy") than simple loading coils, especially when the components and/or construction are of questionable quality!

Quality of design and construction make some commercially made trap dipoles much less "lossy" than others!

CHAPTER FOUR

Ten Meter Dipole

When looking for a way to install an HF dipole in very restricted space, consider operating on the ten-meter band.

During the most active years of a solar cycle, often a simple ten-meter dipole will work wonders. It's easy to put one up because of its relatively small size.

The formula to calculate the length of a center-fed half-wave dipole on 10 meters is:

- **length (meters) = 142.4 / 28.0 MHz = 5.086 meters.**
- **length (feet) = 467.2 / 28.0 MHz = 16 ft. 8 1/4 inches.**

Of course, as you operate in the higher portion of the band, that dipole length obtained by the formula above will be too long. You can either:

- Trim it to resonance (cut & try a few centimeters on each side of the dipole at a time). For example, a 29

MHz dipole will be about 17.5 cm shorter overall than the 28 MHz dipole.

- Or use an antenna tuner. The automatic antenna tuner of some transmitters might have enough capacitive range to tune out the inductive reactance of the dipole as you move up in frequency within the 10 meter amateur radio band.

The Years Of "DX" Ahead

As I write this, the sun's sunspot activity, in cycle number 24, is slowly subsiding. Yet, it will still expel enough energy toward Earth to maintain the "F layer" in good enough shape, at least sometimes, to continue reflecting 10 meter radio signals, hopefully until about 2019.

Even during years of low sunspot activity, 10 meters can sometimes offer "DX openings". The best way to take advantage of them is to have a good 10 meter antenna system and an excellent receiver.

The 10 meter band is my favorite for low power (QRP) DX.

The band is 1.7 MHz wide! That's a big chunk of radio spectrum, as much as all the lower bands put together! Lots of "elbow room". I hope to meet you there for a QSO one day!

CHAPTER FIVE

HF Verticals

A quarter-wave vertical antenna is often the only solution when an amateur radio operator has no room for a dipole, or inverted "V" on HF.

However, it is important to mention that the HF vertical is much less efficient than a dipole because a significant amount of RF energy is lost in the ground system, even when it's enhanced with many 1/4 wave radials. Losses even increase when traps are inserted in the vertical to make it usable on more than one HF band.

Because of the above-mentioned losses, the efficiency of a vertical antenna on HF is typically less than 50% of a full length half-wave dipole's efficiency. This already significant loss of efficiency gets worse when the vertical's physical height is made less than one-quarter wavelength high with the use of loading coils!

1/4 Wave   Vertical

Radials

www.HamRadioSecrets.com

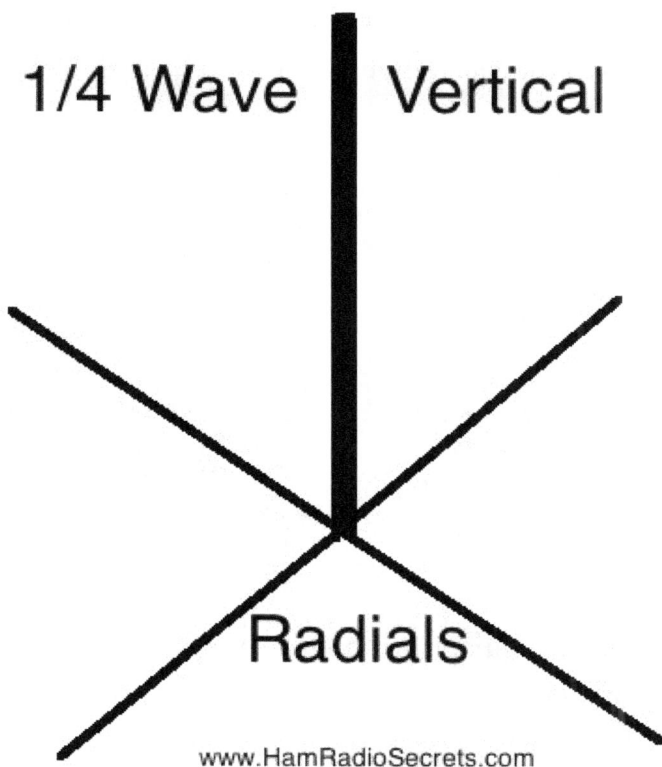

A quarter-wavelength can be obtained with the following formula (approximation suitable for most locations):

- **234 / frequency (MHz) = height of vertical antenna (without traps) in feet**
- **71.25 / frequency (MHz) = height of vertical antenna (without traps) in meters**

To give you an idea of the "size" (height) of a full-length

Page 51

quarter-wave vertical (without traps or loading coils), here are the values for each amateur radio frequency bands (in the Americas):

BAND	FREQ (Mhz)	LOWEST FREQ. (Mhz)	HEIGHT (FT)	HEIGHT (METERS)
160 M	1.8 - 2.0	1.8	130.0	39.6
80 M	3.5 - 4.0	3.5	66.9	20.4
40 M	7.0 - 7.3	7	33.4	10.2
30 M	10.1 - 10.15	10.1	23.2	7.1
20 M	14.0 - 14.35	14	16.7	5.1
17 M	18.068 - 18.168	18.068	13.0	3.9
15 M	21.0 - 21.45	21	11.1	3.4
12 M	24.89 - 24.99	24.89	9.4	2.9
10 M	28.0 - 29.7	28	8.4	2.5

You can see for yourself which full length vertical antenna (without traps or loading coils) is realistically best for you.

More RF Losses

You should also be aware that a physically shorter than normal quarter-wave vertical will offer a much narrower 2:1 SWR bandwidth response than a full size dipole.

This means that, if your transceiver is not equipped with a built-in antenna tuner, or if its tuner cannot handle the complex impedances of the vertical antenna system, then you will have to use an outboard antenna tuner (aka:

"Transmatch"). Unfortunately, most tuners on the market today will cause further RF energy losses because they use "T" networks. You will learn about more efficient tuners in **Book Four** of this series.

Multiband Trap Verticals

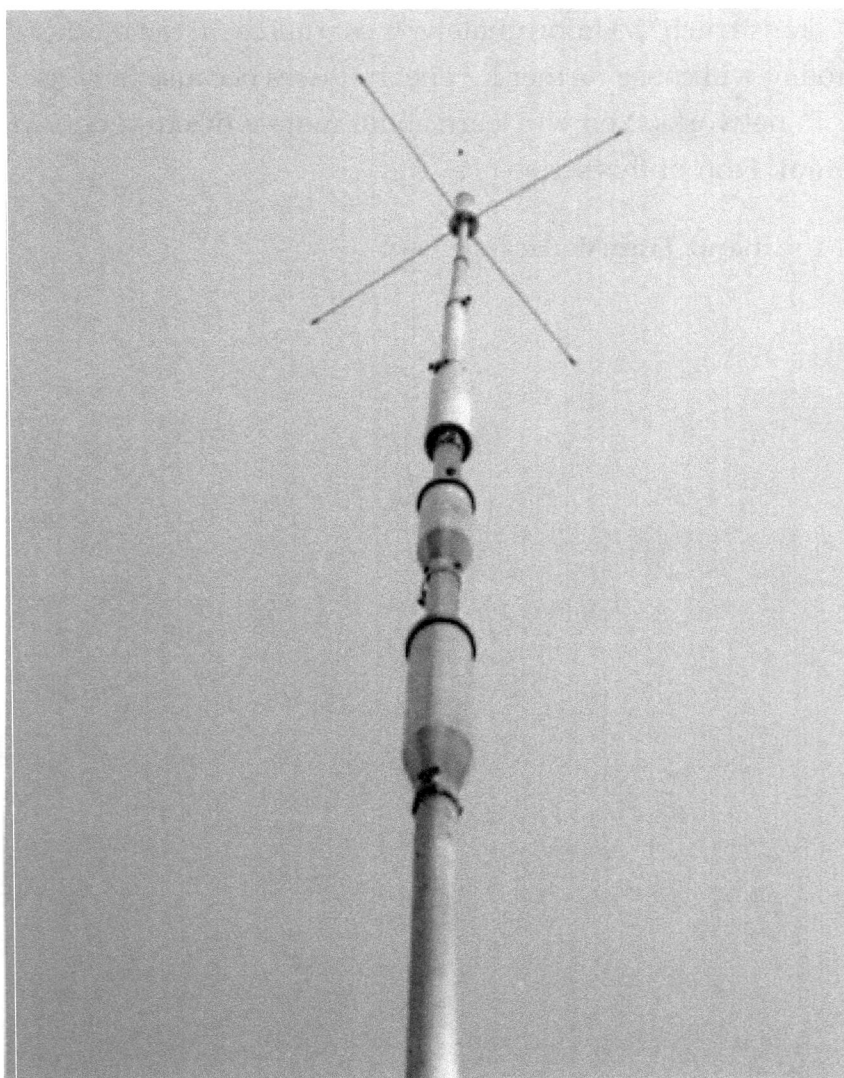

Commercially made vertical antennas, that claim to be designed to perform well on more than one amateur frequency band, will yield **much less than half** the efficiency of dipoles on any band. This is because they must use traps to enable operations on more than one band, as well as loading coils to bring the vertical size down to manageable heights on the lower frequency bands such as 80 and 40 meters.

More Shortcomings

To make matters worse, most amateur radio operators opting for a vertical do not have enough space to install full-size quarter-wave radials on 80 or 40 meters. That is unfortunate because full-size radials would help minimize the severe losses of RF energy in the ground system on those bands.

Promises Promises

If you are stuck with too small a space for anything else, the vertical antenna will be better than no antenna, of course. Just do not expect miracles, no matter what your chosen antenna manufacturer would like you to believe.

In theory, a properly installed **full-size** vertical on HF will have a low angle radiation pattern which will be lower toward the horizon, thus making it suitable for DX, under favorable propagation conditions. But that is if, **and only if**, the vertical is:

- Installed at least one quarter-wavelength (at the

lowest frequency of operation) away from buildings, trees, power lines, and any metal structure greater in size than a quarter-wavelength at the operating frequency.

- Has at least four quarter-wavelength radials connected at its base.
- Has a good RF ground.

However, the radiation pattern of a "loaded" (with coil inductors) vertical antenna will **not** be as low toward the horizon. Physically short verticals, even when resonant at the desired frequency, are not as effective as a full size vertical.

A parting comment on HF verticals for base operation. If you have enough space to install an HF vertical properly, with quarter-wavelength radials as described above, then you might as well install a much more efficient half wave dipole antenna.

Verticals For Mobile Operations

Obviously, the only viable antenna to operate on HF from a vehicle is a vertical. Single band operation is usually the only "relatively viable" solution. In such cases, the antenna will have an inductor (coil) to keep the overall height of the antenna down to something manageable. It is possible to operate on more than one band by introducing traps in the antenna design.

Commercially made mobile trap antennas for HF are available. The information above, on the efficiency of verticals, should give you a good idea of what performance to expect from a mobile HF vertical antenna.

Admittedly, a very inefficient antenna is still better than no antenna.

There is, however, a very effective technique to enhance a mobile antenna's performance when operating stationary, such as in a rest area or a camping ground.

It consists in attaching a "counterpoise" 1/4 wave long insulated wire (on the frequency you want to operate) to the base of the antenna. Then extend it away from your vehicle in the direction you want to make contacts. It can lie on, or just above, ground as a "counterpoise". The improved results will amaze you! Just remember to disconnect and stow away the "counterpoise" wire before you leave.

As a reminder, the formula for a 1/4 wave long radial or "counterpoise" is:

- **234 / frequency (MHz) = length in feet**
- **71.25 / frequency (MHz) = length in meters**

Towers and Directional Antennas

Another way to install an antenna for HF operations, when real estate is at a premium, is to install a trap dipole or multielement directional antenna on top of a tower or mast.

If municipal bylaws will allow it, go for it! But be prepared to pay dearly for the privilege. Towers, directional antennas, rotators cost a great deal of money. Some municipalities will even require that your tower installation be planned by a competent structural engineer, and be installed by professionals, which will add to the overall cost.

You will also have to declare the tower to your home insurer which will increase cost as well.

You can dream, but inquire carefully before you go on a shopping spree!

A final bit of friendly advice: a tower with a beam antenna on top constitutes a very noticeable structure. Make sure you are on good terms with your neighbors.

— 73 —

*** BOOK THREE ***

Homemade
HF Antennas

About e-Book Three

Imagine! You are suddenly getting 59+ signal reports. You proudly answer that you are running "barefoot" and using a homemade HF antenna. You're in amateur radio heaven.

Nothing compares to the intense satisfaction that the amateur radio operator derives from having built a fully functional homemade amateur radio HF antenna. Planning, gathering the parts, assembling, adapting, testing and, finally, making memorable contacts with your homemade antenna are priceless moments in an amateur's life. The rewards are many, as you will see.

This e-book is designed to help you reach that level of satisfaction.

CHAPTER ONE

Antenna Modeling Software

A homemade HF antenna project has the potential to yield *five* rewards:

1. The pleasure of experimenting with concepts.
2. The joy of gathering the parts, building and tweaking to satisfaction.
3. The thrill of going "live" on the air and making the first contacts.
4. The admiration, if not the envy, of fellow amateur radio operators.
5. An economical yet effective HF antenna solution.

First, one of the most enjoyable and rewarding ways to experiment with antenna *concepts* is to use antenna modeling software.

Better Late Than Never

I decided to investigate antenna modeling software because I was tired of the traditional "cut-n-try" method, which was fine during my first quarter of century as an amateur radio operator and antenna enthusiast.

But, now in my late sixties, I feel I need to make more efficient and effective use of my time.

Many consider antenna-modeling programs to be "overkill" for most, if not all, of the needs of the amateur radio antenna experimenter.

Not me, obviously.

The first benefit you get from using a good antenna modeling program is a much better understanding of all the parameters and compromises that go into the design of an antenna system.

Even more importantly, this is how we come learn that there is more to an antenna system than the antenna itself.

Here are my top picks.

Free Software

Here are some of the most popular "no-cost" programs that are available today. To be sure, there **is** a cost to this type of software. Your time! You will have to invest a bit of time up front to learn:

- What the software can do and what its limits are.
- How to find your way around the user interface.
- How to get the most out of any given antenna modeling software.

But the time invested in learning how to design your own antennas will bring very appreciable returns:

- Precious knowledge and know-how.
- The self-satisfaction of having mastered yet another aspect of amateur radio.
- The ability to design *personalized* antennas that will sometimes outperform commercial ones!

NOTE: The following antenna-modeling software run on Windows only.

MININEC

In 2014, Roger, WB0DGF, released an improved version of *MININEC* which he named *miniNEC* v.1.0. The previous version of *MININEC* dated back to 1999!

Roger's version is certainly worth exploring if the above-mentioned software does not meet your requirements and/or expectations. His version of *MININEC* will run in Windows XP, Vista, 7 and 8.

You can download *miniNEC* v.1.0 on WB0DGF's website: http://www.w8io.com/mininec.htm.

EZNEC

The "demo version" of the software is fully functional and more than adequate for most amateur radio antenna-design needs, even if it is limited to subdividing an antenna into only 20 segments. It is very easy to use, compared to other NEC-based software. Oh! And, best of all, it is free.

Once you become familiar with all the capabilities and features of the demo version, you will have acquired the know-how to design more complex antennas. In fact, *EZNEC* is so powerful that you can design complete antenna systems consisting of:

- The antenna.
- Its transmission line.
- Even the impedance transformer (a.k.a. "antenna tuner"), if required.

If a picture is worth a thousand words, a video is worth a

thousand pictures.

Therefore, I produced a video demo
(https://youtu.be/N8Muudm5DdM) to introduce you to
the *EZNEC* v5.0 free demo version. The video may take
some time to load if you do not have a broadband Internet
access. I have had to limit the resolution to reduce the size
of the file, which explains why the final video is "a little
fuzzy".

In the video demo, I use the program to design a 20-meter
dipole which is to be installed 30 feet above ground. The
video will take you through:

1. Initial setup of antenna and environmental
   parameters.
2. Basic antenna-design experimentation.

You can do a lot more with the *EZNEC* v5.0 free demo
version than what I show in the video.

Please note that *EZNEC* v.6.0 has been released since my
video review. The v6.0 demo version has more to offer than
the previous version. I encourage you to give it a try. It is
available at this URL:
http://www.eznec.com/demoinfo.htm.

The documentation is very complete and well written. It is
in the form of Windows Help files. The contents are very

well structured and easy to navigate. You can follow the Help as if it were a tutorial in antenna modeling. You can also skip and jump to specific subtopics using the detailed table of contents.

You will find that the demo version of *EZNEC* v.6.0 is very powerful. On top of being educational, the software will enable you to analyze and design most of the common types of antennas that amateur radio operators use. When you feel comfortable with the program, and like what it can do for you, then you can go for the commercial version which will enable you to explore much more complex antenna systems. You will find more information (below) on the different versions available, and their cost.

AutoEZ

Now, here is a particularly useful utility that enables you to run *EZNEC* in "batch mode". Quoting from Dan Maguire's (AC6LA) website: *"You can run multiple EZNEC test cases while AutoEZ automatically changes one or more variables between runs."* It is available at: http://ac6la.com/autoez.html.

4NEC2

The user interface (UI) and the impressive list of features of the 4NEC2 Antenna Modeler and Optimizer have much improved over the years. It is hard to believe that it is still free.

4NEC2 lets you subdivide your antenna into as many as 1500 wires and/or segments, which means that you can obtain a degree of accuracy, and the ability to model complex antenna systems, which should meet the normal requirements of the amateur radio experimenter. It is available at: http://www.qsl.net/4nec2/.

SMITH CHARTS Demystified

What is this topic doing in the antenna modeling software section? Well, most antenna modeling software packages will display the results of computations on a *Smith Chart*, among other graphical representations. Therefore, it is useful, if not essential, to be able to understand and interpret the results that are displayed on a *Smith Chart*. Here is the clearest explanation of the *Smith Chart* that I have ever come across. It is by Dan Maguire, AC6LA: http://www.ac6la.com/stss.html.

If you still do not understand *Smith Charts*, after this, then I suggest that you refer to the *ARRL Antenna Book*. It devotes an entire chapter to an in-depth description of them.

VE3SQB Software Utilities

His excellent software suite is ideal for those of you who are in a real hurry to get quick but useful antenna solutions, without having to learn too much about what compromises go into the designs, and why.

Al Legary's software suite is as close to "plug & play" antenna modeling software as you will ever get. Neat and simple. My review of VE3SQB's antenna software utilities follows.

Antenna Software
For Amateur Radio Newbies

In spite of the above headline, VE3SQB's antenna software suite is very capable. It is just that, compared to *EZNEC*, *4NEC2*, and the likes, all of VE3SQB's programs are so easy to use.

In addition, his software installs without messing with Windows's registry, which means you can delete the executable files just like any other file. Nice touch! VE3SQB is to be congratulated for his work and for caring for the "non-initiated". The last time I had a look at the collection of programs on his site, I counted 24 executables.

I only cover a small selection of his useful antenna software (below), which should be enough to whet your appetite.

Al admits that he is a "quad" man.

Having had a *GEM Quad* tribander (HF 20M-15M-10M bands) for a few years, back in the late 70s, I can testify to a quad's slight advantages over a tri-bander Yagi with the same number of elements.

My experience has been that a well designed, 3-element tri-band quad has:

- A noticeably broader bandwidth response below a 2:1 SWR than a Yagi.
- A better performance in QSB and marginal conditions than the Yagi.
- A better performance, in receive mode, than a tri-bander Yagi with traps. The quad offers slightly higher gain for the same number of elements.
- A noticeably much quieter performance on receive (picks up much less QRN). The result is that the quad has a better signal to noise ratio than the Yagi.

Al has more to say about "Why Quads"? at the following URL: http://www.ve3sqb.com/whyquads.html.

Quad Design V3

QUAD ANTENNA DESIGN V3 by VE3SQB [EXIT]

- STANDARD HIGH GAIN
- DOUBLE REFLECTOR
- SPECIAL 4 OR 8 ELEMENT

AWG # WIRE SIZE 10

NUMBER OF ELEMENTS 3

INPUT CENTER FREQUENCY IN MHZ

14.1

SET FREQUENCY

CALCULATE/FT _INCHES

CONVERT TO METRIC

MATCHING INFO

WEB LINK   PRINT

	FEET	INCHES
DIRECTORS ARE	68	1.584
THE DRIVEN IS	71	3.319
THE REFLECTOR IS	74	5.053
ELEMENT SPACING FOR		
...........50 OHMS IS	12	6.676
...........75 OHMS IS	14	11.97
...........125 OHMS IS	20	2.755
BOOM LENGTH		
.....FOR 50 OHMS IS	25	1.352
.....FOR 75 OHMS IS	29	11.94
.....FOR 125 OHMS IS	40	5.511
APPROX SPREADER		
HOLE SPACING IS	25	2.446
AFTER THE 2nd DIRECTOR YOU MAY	2	0.527
PROGRESSIVELY REDUCE EACH DIRECTOR LENGTH BY ANOTHER		

BEST USUABLE FREQUENCY RANGE

13.8983   14.1   14.3016

1.5 _____ 1.7

SWR
ESTIMATED FREQUENCY / SWR RATIO

Using Version 3 of his quad design program is very straightforward. You just set the frequency for which you want to design a quad, indicate the number of elements, the wire size, and choose the type of gain you want. Then, hit the "Calculate" button. Presto! All dimensions appear in the table on the right hand side. As with most of his programs, you can choose between English and metric measurements. A graph (bottom right) shows an SWR curve within the "working" frequency range of the quad. This is really straightforward antenna software.

The *"Sky Hopper"*

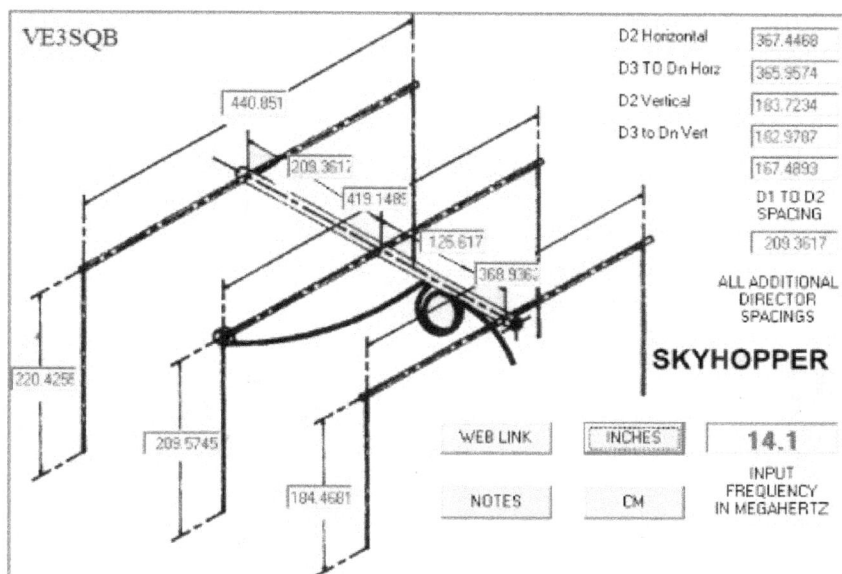

VE3SQB

D2 Horizontal	367.4468
D3 TO Dn Horz	365.9574
D2 Vertical	183.7234
D3 to Dn Vert	182.9787
	167.4893

D1 TO D2 SPACING

209.3617

ALL ADDITIONAL DIRECTOR SPACINGS

440.851
209.361
419.148
125.617
368.936
220.425
209.5745
184.4681

SKYHOPPER

| WEB LINK | INCHES | **14.1** |
| NOTES | CM | INPUT FREQUENCY IN MEGAHERTZ |

Al named this antenna *"Sky Hopper"* because it resembles a grasshopper!

The antenna is essentially a quad with the bottom removed, leaving "U" shaped elements.

Al recommends this antenna for beginners, as it is very "forgiving" of small departures from the dimensions given by the program.

Maximum radiation from the *"Sky Hopper"* is off the corner opposite the feed.

OmniQuad

OMNIQUAD ANTENNA *by* **VE3SQB**

| INPUT FREQUENCY IN MHZ | 14.1 | ACCEPT |

(•) THREE SIDED   ( ) FOUR SIDED

RECOMMENDED MINIMUM WIRE/TUBE

| 10 | **AWG** | 0.102 | **INCH** | 2.59 | **MM** |

CHANGE   CHANGE   CHANGE

COMPUTE

HEIGHT	273.258	**IN**	6940.755	**MM**
SIDE	101.475		2577.467	
GAP	3.679		93.444	

| 300 OHM SERIES MATCH LENGHT | 171.7205 | | 4361.702 | |

WEBSITE   NOTES

The *"OmniQuad"* is omnidirectional. It is very broadband and dimensions are not critical at all, even at VHF and UHF frequencies. Truly a *"fun"* design to experiment with.

Capacitor Designer

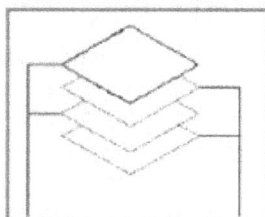

HOMEBREW CAPACITORS
by VE3SQB

Variable capacitors can be made by mounting a plate by means of an adjustable screw.

SELECT

INSULATING MATERIAL

AIR
BAKELITE
ACETATE
GLASS
PLEXIGLASS
POLYETHYLENE

NUMBER OF PLATES   9

SEPARATION OF PLATES   .2

AREA OF PLATES   016

INPUT WAS IN?   INCHES   CM

WEBSITE   *CAPACITANCE IN PICOFARADS IS*   143.36

This is an extremely useful utility program for the experimenter because capacitors suitable for the high voltages present in antenna systems and antenna tuners are hard to find, and very expensive when you do find any.

So now you can make your own, and be able to complete your HF antenna project.

You will find VE3SQB's entire collection of useful antenna software at this URL: http://www.ve3sqb.com/.

LOW COST SOFTWARE

Some things come with a price tag in this world. The antenna design software packing those little extras that you just can't live without is one of those things. Here are a few examples of low-cost, yet very capable, antenna modeling software.

NEC2GO, by Nova Plus Software. Offers more functionalities than *MiniNec*. Available at: http://www.nec2go.com/product.asp. Cost: US$39.95.

MININEC Pro, by Black Cat Systems, is available for both Windows and Mac OS X. Available at: http://www.blackcatsystems.com/software/mininec.html. Cost: US$29.

HIGHER-COST SOFTWARE

Here are two examples, in no particular order:

EZNEC

This is the paid version of the free demo mentioned previously. It costs US$99. The software is also available in (still more expensive) versions for the more advanced experimenter. These are: *EZNEC+* (US$149), *EZNEC Pro/2* (US$525) and *EZNEC Pro/4* (US$675). Available at: http://eznec.com/eznec.htm. Note that he *Pro/2* and *Pro/4* versions require the purchase of a *NEC* license. Its cost will range from $300 to $1,500, depending on which type of use you will make of *NEC*. See details here: https://ipo.llnl.gov/technologies/nec.

NEC2 and *NEC4* modeling files for *EZNEC*, compiled by L.B. Cebik, are available on the *AntenneX* website: http://guests.antennex.com/rooms/software/index.htm. Y will need to register as a guest. It's free. Once you receive your password, login, and:

1. Go to the *AntenneX* Guest rooms;
2. Click on the link labeled *Software*, located in the left-hand column, under Special Sections/Programs/Models.

PCAAD 7.0

Personal Computer Aided Antenna Design (PCAAD) is by Antenna Design Associates, Inc. Quoting from their website: *"PCAAD 7.0 is intended for use by engineers, students, and researchers who need quick solutions to common antenna design and analysis problems. PCAAD 7.0 is **not** downloadable. It is supplied on a CD, along with a 120-page printed user manual."* Cost: US$549, plus shipping. Available at: http://antennadesignassociates.com/pcaad7.htm.

SOFTWARE FOR MAC OS X

The following antenna modeling or design software is available for those of us who prefer to work in the Mac OS X environment.

MININEC Pro, by Black Cat Systems (US$29). Available at: http://www.blackcatsystems.com/software/mininec.html.

CocoaNEC 2.0, by Kok Chen, W7AY. Available at: http://www.w7ay.net/site/Applications/cocoaNEC/inde:

RF Toolbox, by Black Cat Systems. Available at: http://www.blackcatsystems.com/software/antenna.html.

CHAPTER TWO

Antenna Analyzers

For those of us who do not like to deal too much with the theoretical side of things, and enjoy keeping things simple, there is an easier way to build antennas than to start with antenna modeling.

You can start with the basic formula to calculate the length of HF antenna elements for the desired frequency of operation. Then experiment with different configurations (prototypes) using an antenna analyzer as a guide.

If you can afford one, an antenna analyzer will greatly reduce the traditional and often frustrating trial and error method of building a homemade antenna. In fact, it will probably encourage you to experiment with more than one type of antenna. This is desirable, of course, because it will greatly increase the level of satisfaction you will derive from the amateur radio hobby.

The cost of a unit will range from a few hundred dollars to a few thousand, depending on its degree of sophistication and quality of craftsmanship. A unit to analyze HF antennas will cost in the neighbourhood of $US200. But not all are made equal. Some are not worth this price, while others are a bargain in comparison.

I will cover antenna analyzers in more detail in *Book Four*.

CHAPTER THREE

An Antenna
For 160 Meters

I chose this homemade HF antenna project to illustrate that it is possible to obtain excellent results, even when you have to bend the rules somewhat to beat the odds.

My antenna for the 160-meter band is the result of many considerations, inner debates and tests. The *ARRL Antenna Book* was very useful in helping me understand the variables. I used an antenna analyser to help me find the configuration that presented the best characteristics.

In the end, the love of experimenting with antennas played a critical part in the successful outcome of my homemade antenna project.

The Compromises

Every antenna is a set of compromises. This one is no exception. I had to take into account:

- Available horizontal space.
- Available supports (tower and trees).

Many amateur radio operators, when short on space, will think of a quarter wave vertical. In this case, a 1/4 wave vertical was out of the question, being just under 40 meters in height (130 feet) on 160. Granted, I could have considered a top "loaded" vertical, with a top coil (inductance) to make up for a short physical height.

But, I do not like verticals, "loaded" or not. They may be good enough for transmitting energy (+/-) but they are poor receiving antennas, being so short relative to the desired wavelength.

Furthermore, quarter wave verticals just "bleed" too much precious RF energy in the ground system for my taste.

So I had to choose some form of "roughly balanced" horizontally deployed antenna.

Choosing The Lesser Of Many Evils

After playing with many odd configurations, I finally settled on a dipole for its inherently balanced and higher efficiency in capturing and radiating energy. But I still had to fit (squeeze) the 160M dipole into the boundaries of my lot.

So, I bent the rules, and the antenna, somewhat.

The result is a "hybrid" dipole or "ungrounded" inverted "L" vertical antenna with a counterpoise, depending on your point of view.

The feed point is just outside my shack window, about 1.8 meters (6 feet) above ground.

I feed the antenna with a short length (less than 16 feet) of RG-8X through a RemoteBalun® (current balun) made by Radio Works. The balun prevents RF energy from leaking back into the shack via the coax shield.

One side of the dipole goes up from near ground level to about 14 meters (45 feet), then out 25 meters (82 feet) toward the front of my lot. This portion becomes an inverted "L" un-grounded quarter-wave vertical, 39 meters in length, made up of #14 stranded (7x22) copper-clad antenna wire.

The other side of the dipole is a quarter wave "floating" counterpoise for the inverted "L" vertical portion. It drops down toward the back of the house, and is supported by tree branches about 6 feet (1.8 meters) above ground. I use #14 insulated multi-strand wire. It is thus roughly deployed in the opposite direction from the inverted "L". I recommend that you seal the end of the wire carefully if it is liable to touch ground or wet snow.

Antenna System Schematic View

VE2DPE
160 Meter
"Hybrid Dipole"
Antenna System

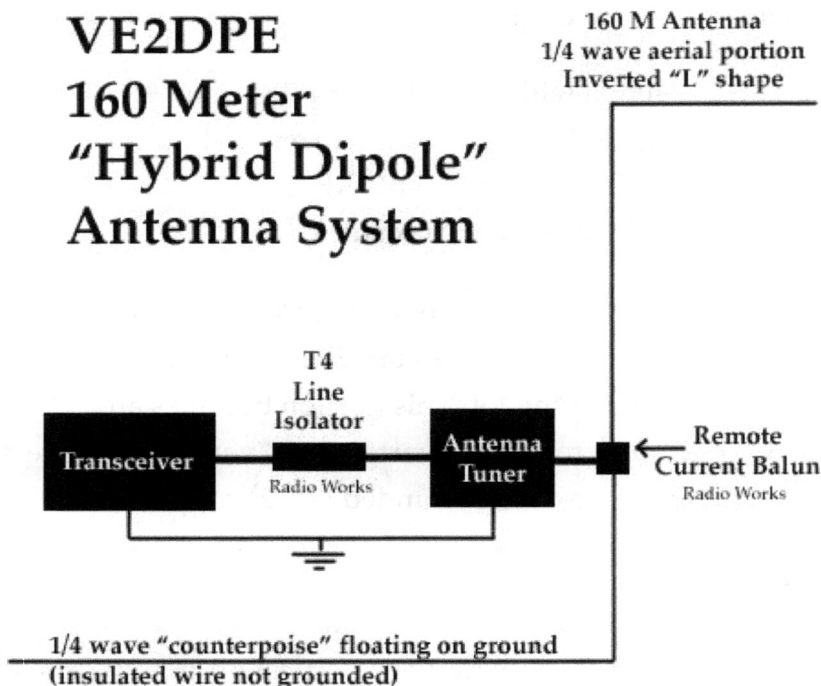

160 M Antenna
1/4 wave aerial portion
Inverted "L" shape

T4
Line
Isolator

Transceiver

Radio Works

Antenna
Tuner

Remote
Current Balun
Radio Works

1/4 wave "counterpoise" floating on ground
(insulated wire not grounded)

www.HamRadioSecrets.com

Overhead View

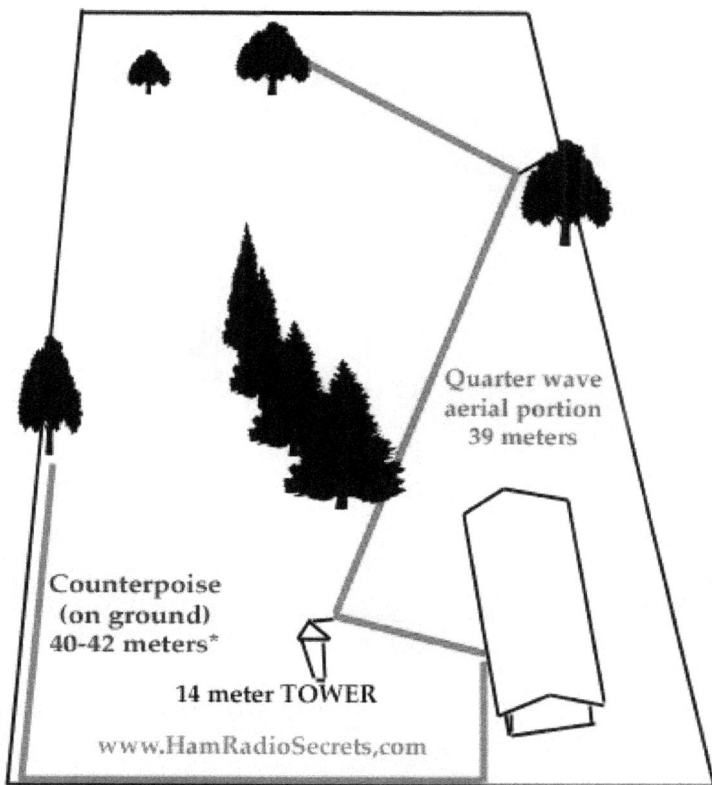

Quarter wave
aerial portion
39 meters

Counterpoise
(on ground)
40-42 meters*

14 meter TOWER

www.HamRadioSecrets.com

View From Back of House

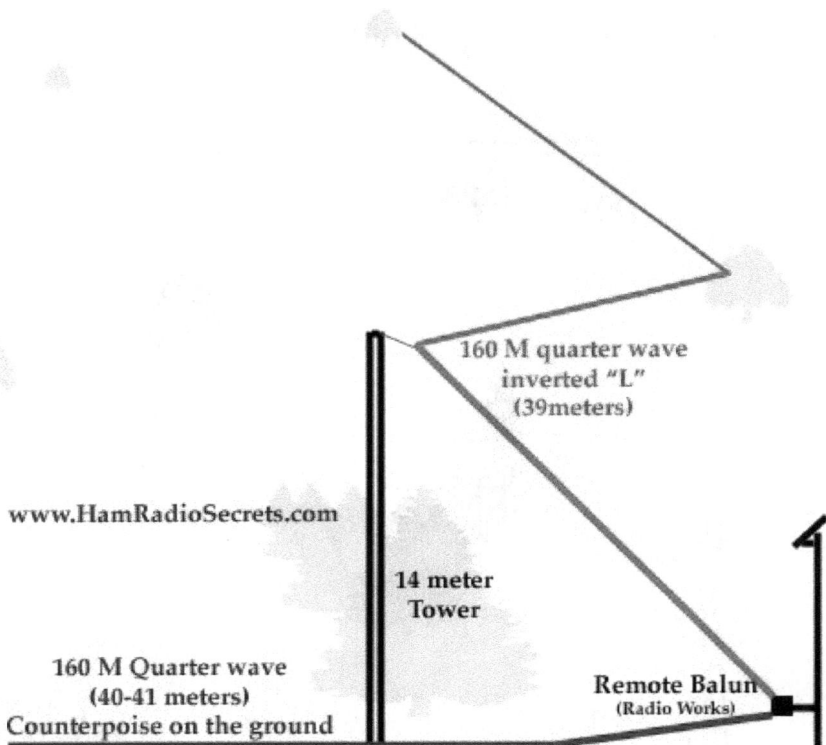

160 M quarter wave
inverted "L"
(39meters)

www.HamRadioSecrets.com

14 meter
Tower

160 M Quarter wave
(40-41 meters)
Counterpoise on the ground

Remote Balun
(Radio Works)

If you have trouble getting your antenna tuner to deal with the reactance of this antenna, you can try cutting back on the length of the counterpoise, about ten (10) centimeters at a time, until you succeed.

The Results
Are What Counts

As you can see, this amateur radio antenna system for 160 meters is "unorthodox", to say the least. It is the best set of compromises I have been able to devise, after many experiments.

It works like a charm, for both local and DX. Here is what Chris, K5LH, had to say about it:

"The antenna keeps producing impressive contacts, is easily erected and taken down, and combines simplicity with remarkable performance. It is definitely a keeper for me. Best wishes and 73".

CHAPTER FOUR

Wire Antenna Tips

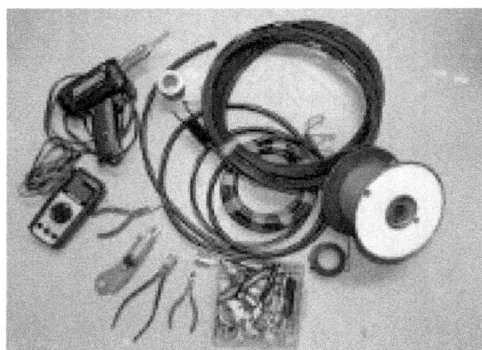

The amateur radio wire antenna is the most common form of antenna used by amateur radio operators. Here is what aspiring and novice amateur radio operators should know about what makes them reliable.

The main reasons for its popularity are that it is affordable and relatively easy to install. In many cases, a wire antenna may be the best solution (set of compromises) for your specific location, especially when you build it yourself.

Cost Aspect

The amateur radio wire antenna is particularly economical when it is homemade. There are a few reasons for this:

- You can use the parts you have in your "spare parts box".
- You can buy parts at flea markets and hamfests for next to nothing (compared to buying new).
- You can use various non-conducting materials as insulators.
- You can make (otherwise expensive) current chokes by coiling your coaxial transmission line at appropriate locations. (Details were given in Book One of this series).
- You learn a lot about antennas in the process.

Wire Types

Before selecting a given wire type for your ham radio wire antenna, you should take the following recommendations into consideration:

- **Antenna length:** obviously, a 160-meter dipole will require much stronger wire (#14 Stranded (7x22) Copper-Clad Steel) than a 10 meter dipole (where

#18 Stranded (7x22) Hard-drawn copper will suffice).

- **Solidity of supports:** a dipole strung between two trees will require special considerations because trees sway in the wind, and will exert tension. It is best to have the antenna rope go through a pulley, at least at one end, and hang a suitable weight at the end of the rope to provide just enough tension to prevent the dipole from sagging too much. Nevertheless, I recommend using #13 19-strand VariFlex® Insulated wire. A dipole, between two solid buildings, will not require you to compensate for variations in tension.

- **Wind:** a notoriously windy location will require stronger wire than a sheltered location.

- **Ice storms:** go for the strongest wire you can afford if you live in an area where an ice storm is likely to occur.

- **Salty air:** you should use wire with a polyethylene or PVC jacket to prevent corrosion from weakening your antenna wire.

- **Heavy industrial pollution:** the recommendation is the same as for a salty air environment. Use jacketed wire.

- **"Stealthiness":** use strong and very thin wire (such as #26 VariFlex® Insulated 19 strand) with sky-colored button insulators, and your antenna will be nearly invisible.

- **Quads, Wire Beams:** I would recommend Flex-Weave® wire by RF Davis. They offer #14 (168 strands) and #12 (259 strands!), bare or insulated (PVC or polyethylene).

- **Portable antennas:** strong and very thin wire such as #26 19 strand VariFlex® Insulated is perfect for this application. Light, durable and very flexible.

Insulators

Center insulator: If you choose not to use a balun at the feed point of an amateur radio balanced antenna, I recommend a glass-filled ABS insulator with built-in drip lip around the SO-239 connector. It comes with a hole on top to anchor inverted-Vs. Also, you might consider the excellent, but slightly more expensive, DELTA-C Center Insulator by Alpha Delta Communications.

End insulators: For long-lasting trouble-free service, I recommend the Delta-CIN End Insulator — especially if you operate at high power. The Delta-C components are molded of high impact UV and RF resistant materials called *Deltalloy®*.

Of course, "generic" ceramic insulators will continue to serve you well in most cases. Even plastic insulators are used for portable QRP work because they are light weight and offer good enough insulation at such low power levels.

Rope Types

When old army surplus ceramic insulators were abundantly available, many of us would use clothesline. It is quite strong and usually fairly UV resistant. Unfortunately, it is too "fat" to fit into the new center and end insulators recommended above.

But big (heavy) ceramic insulators have become harder to find, as well as (non-conducting) clothesline rope.

Dacron® type rope has come to our rescue. It is extremely strong as well as UV and abrasion resistant. Use the 3/8" size for the center insulator and pulley and the 3/16" to tie the end insulators to end supports.

Take down and carefully inspect all parts of your ham radio wire antenna at least once a year for damage.

I do get much better results than any commercially built

wire antenna for 160 meters with my homemade wire antenna. But I must admit that I have not yet tried the Carolina Windom® 160.

Commercially Built Wire Antennas

Commercially manufactured wire antennas do have advantages. They are:

- Generally made with quality, durable parts;
- Designed and pre-tuned for best performance (although admittedly under ideal conditions).

However, expect somewhat different performance results if you have to deviate significantly from the installation instructions.

For example, I have been using a Carolina Windom® for 80-10 meters for years. Even if it is not installed quite as high as recommended (especially the longer leg), it remains the best commercially made multi-band (no traps) single wire antenna that I have ever used.

Performance Aspect

Will the homemade wire antenna perform as well or better than a commercially built wire antenna of the same type? Often it will when you design it to fit your specific surroundings and personal preferences.

Let's face it. Most amateur radio operators buying a commercially made amateur radio wire antenna will be reluctant to modify it. You are likely among them. Therefore, look *closely* at the installation specifications *before buying*. This means that you will have to get a copy of the installation manual. Most responsible manufacturers make them available online. If you find that you will **not** be able to meet the installation requirements, then try a homemade solution instead. You stand a good chance of getting better results than a poorly installed commercially-made wire antenna.

CHAPTER FIVE

"Stealth" HF Antennas

In *Book Two*, I promised to provide techniques to design, build and install a "low-profile" — sometimes called "stealth" — outdoor HF antenna when city bylaws (or uncooperative neighbors) will not otherwise allow one. When that is not possible, I will also give you indoor antenna solutions, with its pros and cons.

Low-profile Outdoor HF Antennas

Remember: Bees may look clumsy, yet they *can* fly! Your idea, even if it seems far-fetched, could too. You will not know until you try.

Test an idea before rejecting it as useless. Your first low profile HF antenna idea may not work as hoped, but a different configuration might. In fact, once you get your brain in high gear, other ideas will almost certainly come.

Building homemade antennas is much like writing nonfiction. You have to get started if you are going to get it done.

My method is to start by jotting down all the parameters that you know:

- Potential locations for a low profile antenna
- Available space (dimensions) at each location
- Available hardware (antenna wire, insulators, transmission line). Use what you have before buying.

Here are a few installation ideas:

- A 1/4 wave vertical, hung down from a high tree branch, hugging the tree trunk (or tall wooden structure), with one or more 1/4 wave radials above ground (counterpoise), lying on the ground or buried. See recommendations, in the chapter on transmission lines, on how to protect coax lying on the ground or buried.

- An inverted "L" 1/4 wave vertical element, with a counterpoise down to just above ground, like my 160M example previously mentioned. Run the top element from the edge of the rooftop to a tree or other vertical support using thin, light-colored wire and grey insulators. The counterpoise portion of the antenna can be stretched out, in the general oposite direction, down to no less than 6 feet above ground.

- A solution often overlooked: install a mobile (vertical) multiband HF antenna on your vehicle and run a coax to your house (buried if you must absolutely hide it from view). Most city bylaws cannot prohibit a mobile antenna. Just do not forget to unplug the coaxial transmission line before driving off. However, a mobile HF antenna is not a "low-profile" antenna. A neighbor might complain of real or imagined interference at the mere sight of it.

- Are you allowed to have a flagpole on your property? Disguise a ground-based HF vertical antenna into a flagpole by slipping a white PVC pipe over it.

- Is your house exterior all brick or covered with non-conductive siding? Run a wire loop around your house, snug against the wall, as high up as you can. Use insulated wire with a color to match that of your outside walls as closely as possible.

- Do you have a wooden fence? String an insulated wire dipole along the top.

- Do you live in an apartment building? How about an upside-down vertical? Drop a 1/4 wave wire down along the outside wall from a window and run a counterpoise 1/4 wave wire on the floor of your apartment. Try to avoid hanging the wire in front of your neighbours' windows on the lower floors.

- As a last resort, you could try a dipole lying flat on the ground, using insulated wire and sealed connections. This is a poor performer because the ground reflection will cancel most of the signal. But, if the soil in your area is very sandy and dry, you might find that this dipole will perform better than anything else you might try indoors.

WARNING! In all cases where the antenna is close to human activity, put safety considerations first in your HF antenna project planning. See next section on indoor antennas for details.

Indoor HF Antennas

Operating an indoor HF antenna is not without potential drawbacks, and even dangers. You must be acutely aware of them, and of the measures required to eliminate them.

Safety First!

There is a very real danger of emitting RF energy exceeding the safe levels for humans. I highly recommend limiting your transmitter's output power to 100 watts or less, and installing the antenna as far away from human activity as possible. As you will see in a moment, you have to use much less power on ten meters than on twenty meters to remain on the safe side. I strongly suggest that you familiarize yourself with RF radiation and Electromagnetic Field Safety. To do so, please:

- Read this PDF file:
  http://www.arrl.org/files/file/Technology/tis/inf
- Or visit the following website for more information:
  http://www.arrl.org/rf-radiation-and-electromagnetic-field-safety.

If you operate a 20-meter dipole (14.0-14.35 MHz) in the attic, and feed it 100 watts of RF, your antenna will generate an estimated 0.364 mW/cm. sq, 10 feet away from the center of the antenna — which lies well within the limit for maximum permissible exposure (MPE) of 92 mW/cm. sq at 14 MHz. (The MPE limit for any frequency within the allocated amateur radio HF bands can be calculated using the formula 180/freq(MHz)squared — Ref: ARRL Antenna Book, 21st Ed. 2007, Chapter 1, page 20). If anyone is liable to be closer than 10 feet away from the antenna, then you will have to reduce power to avoid exceeding the MPE limit.

However, the MPE limit drops as you increase your operating frequency. The RF power density generated by a 10 meter dipole in the attic will be 0.364 mW/cm. sq — far exceeding the MPE limit of 0.2346 mW/cm. sq, ten feet away, at that frequency. Therefore, you will have to reduce transmitting power to 50 watts or so to avoid exceeding the MPE limit (the RF power density at 50 watts with this antenna will be 0.182 mW/cm. sq).

Here is a link to a handy RF Safety Calculator to help you decide what type of indoor antenna to install and where. *Use it as a general guide only!*
http://hintlink.com/power_density.htm.

Remember: Operate an HF transmitting antenna indoors at your own risk!

Radio Frequency Interference (RFI)

Radio frequency emissions also create a very real possibility of causing radio frequency interference (RFI) to delicate electronic devices, yours and your neighbor's — disrupting their normal functioning, or worse, damaging some of them permanently. With the trend toward *"Internet of Things"* (IoT), RFI will become an increasing concern. If you encounter this problem, the solutions are:

- Reduce transmitter power to 5 watts or less.
- Relocate the antenna further away from the affected devices. Be aware that, in doing so, you may just be transferring the interference to a different set of devices;
- Revert to an outdoor low profile antenna.

Other Health Hazards

You must eliminate the possibility that someone in the household will come in contact with the antenna. Even at 5 watts of RF power, some antenna configurations can develop very high voltage "hot points".

Standing Wave Ratio (SWR)

RF feedback in the shack, due to voltage standing waves (SWR) on the transmission line, can cause your receiver or transceiver to malfunction. Even worse, you can experience RF stinging on your fingers or even your lips when speaking too close to the microphone. A choke balun where the transmission line connects to the antenna should eliminate this possibility. You can use a commercially made one, or make your own choke by winding (coiling) the coaxial transmission line. See Book One of this series for details.

Indoor Antenna Location

After having read the above caveats, if you still have no other choice than to resort to an indoor HF antenna, then here are your options:

- In the attic (if you are fortunate enough to have one).
- On the ceiling of one or more rooms of your house or apartment.
- Along the baseboards of one or more rooms, located in the top story of your house.

In all cases, you will need to use a very efficient antenna system tuner. Most popular low-cost tuners waste so much RF energy in heat that they could serve as dummy loads! You will find details of what types of tuners are more efficient than others, and why, in *Book Four* of this series on *Amateur Radio HF Antennas*.

You should use insulated wire. Avoid, as much as possible, running antenna wire parallel to house wiring and long metallic objects such as eavestroughs (just outside, along the edge of your roof) and flashing (also present along the edge of the roof and along angles in the roof surface).

Indoor HF Antenna Configuration Ideas

The diagram below shows an overhead view of a multiband "umbrella frame" HF dipole, with its branches strung between rafters in the attic, and the wires parallel to the ceiling of rooms below, somewhat like a horizontal cobweb.

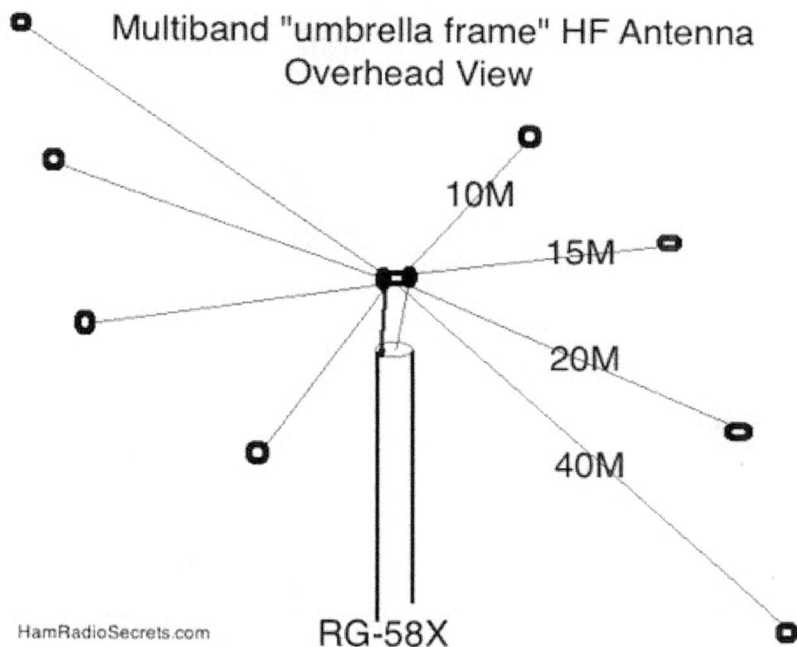

Multiband "umbrella frame" HF Antenna
Overhead View

10M
15M
20M
40M

HamRadioSecrets.com

RG-58X

The next diagram shows an overhead view of a dipole installed along the outer perimeter of the ceiling (or attic) of a single-story house.

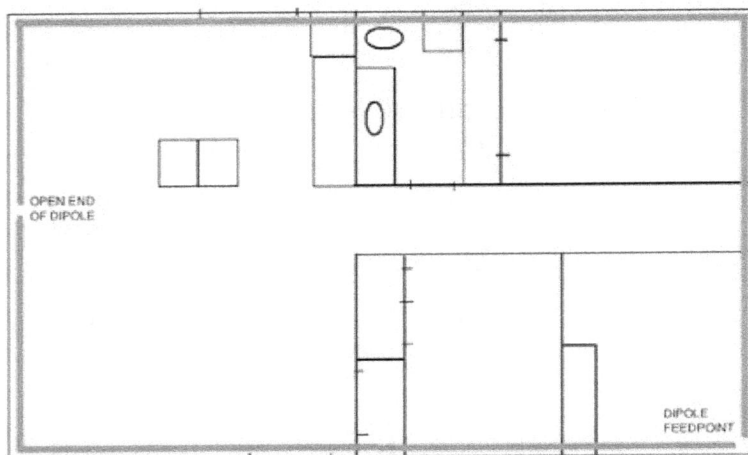

OPEN END
OF DIPOLE

DIPOLE
FEEDPOINT

HamRadioSecrets.com

If the house has two stories, the dipole should be installed in the outer perimeter of the top floor ceiling or attic.

As an alternative, you could install (in the attic) two multiband mobile HF antennas, feed point to feed point, each becoming one half of a dipole.

CHAPTER SIX

Transmission Lines
And Connectors

No antenna *system* is complete without its transmission line. The most appropriate and easiest to use in the case of a low profile HF antenna is the coaxial cable. The following are all suitable for their small diameter and power handling capabilities. Where I stress the difference is in loss (in decibels per 100 feet) because this is a critical parameter in receive mode, especially in the case of low-profile antennas, which are usually installed in less than ideal locations and configurations.

RG-174

The most unobtrusive is RG-174 (only 0.11 in. diam.). But it is lossy: 3.3 DB per 100 feet at 10 MHz, not counting additional losses if you experience to standing waves due to mismatched impedance between the coax and the antenna. You may choose to live with that on transmit mode, as some power will still reach the antenna. But in receive mode, it is going to "strangle" a lot of incoming signals, except the strongest, unless you can limit the length of the transmission line to much, much less than 100 feet.

RG-58

A little thicker at 0.195 in. diam. than RG-174 and less lossy at 1.0 DB per 100 feet at 10 MHz. It has a solid #20 center conductor.

RG-58A

It is the same size as the RG-58 but its #20 center conductor is multi-strand Flex which increases loss a little to somewhere between 1.3 and 1.5 DB per 100 feet, depending on the quality. Quite acceptable, especially for short lengths.

RG-8X

I would recommend RG-8X for its superior outer shield. It is thicker than all the above at 0.242 in. in diameter. Its loss is 0.7-0.8 DB per 100 feet with the #15 solid center conductor (0.9 DB if #16 Flex center conductor).

If you have to bury, or lay any of the above mentioned cables on the ground in a permanent fashion, I suggest slipping them inside a good quality rubber garden hose. Ensure that the hose is well sealed at both ends, and at any hose junction in between. Inspect once a year for moisture seepage. I suggest wrapping a couple of layers of Saran Wrap on hose junctions, then tightly wrap with two layers of electrical tape. Use Coax Seal® to weatherproof the coaxial connector(s) outdoors.

Ladder Line

You can also feed a homemade HF antenna with 450 ohm ladder line or 300 ohm twin lead. However, their characteristic impedance will require some form of impedance transformation down to the nominal 50 ohm of your transceiver antenna connection.

Ideally, to minimize losses in the transformation, you should use a good quality antenna tuner for that purpose. More on these in *Book Four*.

It is not as easy to run a ladder line through walls to your operating position. Therefore, it is best to have the tuner outdoors, and have a length of coax through to your transceiver. Outdoor tuners add to cost and complexity.

Alternatively, you can have a current balun just outside your window. You then connect the ladder line to it outdoors, and run a short length of coax to your rig indoors. This is what I did to feed the second version of my homemade 160M antenna. I used a LDG RBA-1:1 Current Balun for the purpose.

I chose to do it this way to benefit from the extremely low-loss characteristics of the ladder line. Very high SWR occurs along it when I operate on some bands but loss is still minimal. This is particularly important to me when I operate in QRP mode (five watts or less). I want to be able to hear the faint ones in the background, as well as being able to get my signal out to them.

Note that the ladder line is just outside my window. I was getting RF feedback in the shack. To solve the problem, I simply installed clamp-on ferrite chokes on the coax, just before it enters the house. Three of them finally stopped the RF from leaking into the room.

Connectors

I use PL-259 (83-1SP) connectors with adapters (reducers) to fit snugly over small coaxial cables: a UG176/U reducer for RG-8X coax, and a UG-175/U for RG-58 or RG-58A coax.

If you want to have a PL-259 at the end of a RG-174, you will have to use a Lands Precision 8735 connector, or its equivalent. Alternatively, you can install a BNC-male connector at the end of your RG-174 coax, then use a BNC-Female to PL-259 adapter (RFA-8312). However, this latter solution will add a little extra loss to the already lossy RG-174.

CHAPTER SEVEN

Antenna Reference Books

The designing and building of homemade antennas is a vast subject area, and it has become a hobby within a hobby for many of us. Much has been written on the subject. Some of it cannot be ignored. But, before I show you my list of favorite antenna books, allow me to ask:

Who "Brews" The Best Homemade HF Antennas?

I believe that *low power* (QRP) enthusiasts design and build some of the most efficient and effective homemade amateur radio antennas.

Why?

Because they have to extract every micro-volt of RF possible from the radio waves to hear other QRP amateur radio stations. Also, when a QRP operator puts out **5 watts of RF or less** into an antenna system, s/he wants most of it to get out and produce contacts.

Every successful QRP operator knows that her/his success is due to two essential elements:

- Mastering QRP operating techniques.
- Using a well-designed antenna system.

For more on QRP and its antennas, here is a good reference that I recommend without hesitation.

Low Power Communication — The Art and Science of QRP, by Rich Arland, W3OSS, 3rd Edition, ARRL Publication 2007 ISBN: 0-87259-104-2.

ARRL Books

Solid reference books by the American Radio Relay League:

The *ARRL Antenna Book*, 21st Edition. ISBN: 0-87259-987-6

I strongly recommend that you read Chapter 4 of the ARRL Antenna Handbook. The chapter is dedicated to "*Antenna Modeling & System Planning*". It will help you understand the antenna modeling process, and evaluate the antenna modeling software available.

The ARRL Handbook

Of course, the key reference — *ARRL Handbook for Radio Communications* — contains all the technical details to help you understand how antennas work, reducing the time you spend on trial-and-error. A revised edition is released each year.

Note: if you are a US citizen, I encourage you to buy ARRL publications directly from the ARRL, preferably as a member supporting the ARRL. Outside the USA, you might get a better deal by buying from Amazon or some other online discount outlet, rather than by ordering directly from the ARRL.

Visit this URL for more ARRL books on antennas:
http://www.arrl.org/shop/Antennas/.

RSGB Books

Building Successful HF Antennas, by Peter Dodd, G3LDO, RSGB Publication. ISBN 9781-9050-8643-6.

Visit this URL for more RSGB books on antennas:
http://www.rsgbshop.org/acatalog/Online_Catalogue_Ar

Periodicals

The only periodical that I can recommend is *AntenneX*, as it is exclusively dedicated to antennas and covers all aspects.

AntenneX carries articles for every level of expertise, from the beginner to the engineer. Admittedly, many are aimed at the more experienced and technically inclined amateur radio operators. It is available at:
http://www.antennex.com.

— 73 —

*** BOOK FOUR ***

HF Antenna
Accessories

About This e-Book

This fourth e-book in the series on Amateur Radio HF Antennas covers essential, and often ill-understood HF antenna accessories.

Lightning on the horizon? You are not worried ... because you installed the lightning protection devices described in this e-book, and you have taken the preventive measures as prescribed.

But that's not all. Your HF antenna is now transmitting more RF power, and your receiver is getting more milliwatts from your antenna system, because you learned what types of antenna tuners to avoid. You are now using an efficient tuner, and you even know how it works.

Your antenna tuner will likely have its own SWR/PWR meter. Notwithstanding, this e-book will help you decide if you still need an outboard one, and which meter will best meet your needs.

Furthermore, armed with the valuable information you read here, you finally acquired the antenna analyzer to satisfy the expectations of the antenna experimenter in you.

Emboldened by your recent successful acquisitions, you now have the antenna tower you were dreaming of. You heeded the recommendations this e-book provided and, thus, avoided all the pitfalls that await the unwary.

Finally, an e-book series on HF antennas would not be complete without a word on HF signal propagation. It outlines the software and online services available today which enable you to take full advantage of band openings.

CHAPTER ONE

Lightning Protection

Before lightning strikes, here are four measures that are necessary to shield yourself, your equipment and your loved ones from potentially deadly lightning strikes.

Preventive Measure #1

Implement recognized and proven antenna system and station grounding measures and procedures.

Preventive Measure #2

Install lightning arrestors, as prescribed by recognized experts, and manufactured by reputable firms. You will need to protect:

- Every coax cable coming into the radio room;

- Every control cable to directional antenna rotors, switching control box(es), etc. See the list of lightning arrester manufacturers below.

Preventive Measure #3

Use of a single point ground is critical. See diagram.

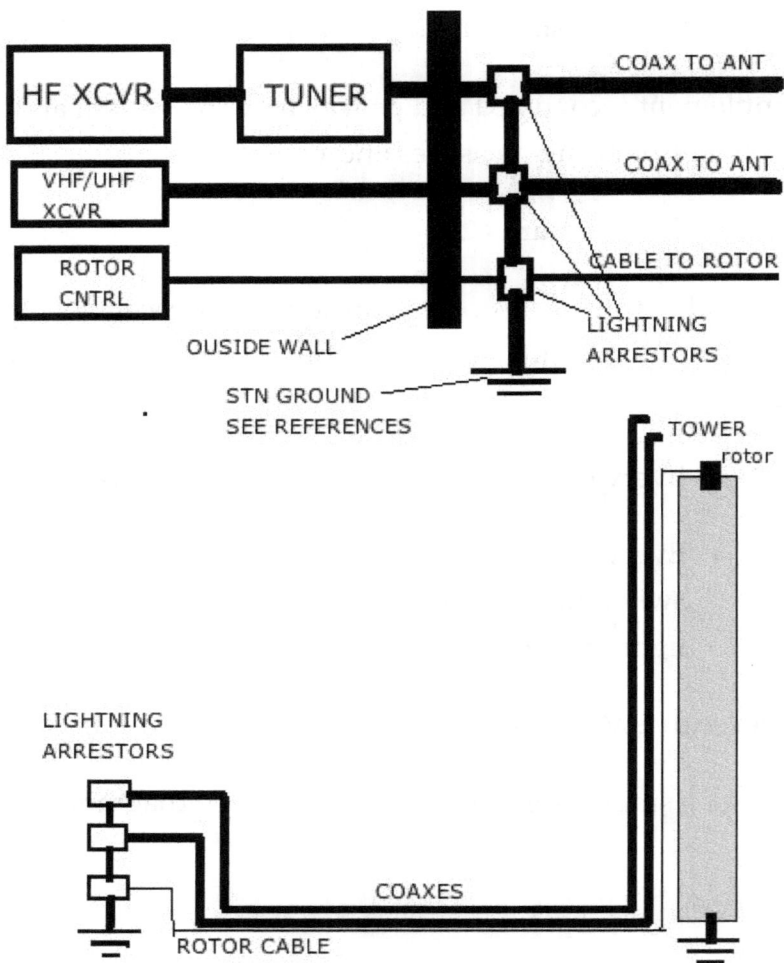

HF XCVR — TUNER — COAX TO ANT

VHF/UHF XCVR — COAX TO ANT

ROTOR CNTRL — CABLE TO ROTOR

OUSIDE WALL

LIGHTNING ARRESTORS

STN GROUND
SEE REFERENCES

TOWER
rotor

LIGHTNING
ARRESTORS

COAXES

ROTOR CABLE

Preventive Measure #4

Always be on the lookout for stormy weather. Use trustworthy sources of weather forecasts for your area such as:

- **United States**

- NOAA Weather Radio (1.usa.gov/1ZBNjwH).
- National Weather Service (www.weather.gov/).

- **Canada**

- WeatherRadio Canada (bit.ly/1Ki3JxV).
- Government of Canada Weather
  http://weather.gc.ca/mainmenu/weather_men
  http://weather.gc.ca/warnings/index_e.html

As soon as the forecast mentions stormy weather is in the offing, prepare yourself.

1. Begin monitoring public storm warnings and alerts.

2. Begin monitoring websites which display the location of lightning activity, and how it is evolving (see below).

3. Begin monitoring your lightning detection system (see below).

Monitoring For Lightning
Weather Radio

A good way to monitor for alerts is to use a Weather Radio

Receiver with S.A.M.E. (Specific Area Message Encoding) receiving capability. Choose one which will provide both audible and visible alerts. Leave it on (plugged in) and within sight or earshot until all alerts and warnings for your area have been cancelled. Always have fully charged batteries at the ready in case electrical power fails.

I use an Oregon Scientific Desktop Emergency Alert Radio Model WR608 to monitor the weather forecast, warnings, watches and advisories. It serves me well both at home, and when we go on long camping trips. We sometimes travel out of range of any weather radio transmitting sites. In such circumstances, the WR608 allows me to plug an outdoor antenna that I install outside our travel trailer when we are stationary. You can see the antenna jack on the right-hand side of the receiver.

The WR608 automatically selects the channel with the strongest signal. However, that may not be the station you need! Therefore, it is wise to bring along a list of all the weather radio transmitting sites. Armed with that, you can select (tune in) the one you need. Usually, that will be the one closest and upstream from your location.

Here is where you can get the complete list of weather radio transmitting stations:

- In the **United States**: 1.usa.gov/1ZBNjwH
- In **Canada**: bit.ly/1QjbSZh

My weather radio receiver is not the most full-featured one. The best radios are those:

- able to receive local AM or FM stations, while still monitoring weather alerts for your area (S.A.M.E.);

- able to use multiple power sources: hand crank generator, solar power, batteries, and main electrical power.

Below is a partial list of dependable full-featured emergency radios:

- **Etón** American Red Cross FRX3
  (http://etoncorp.com/en/productdisplay/frx3-
  american-red-cross).

- **Epica** Emergency Radio
  (http://amzn.to/1ZFb9rr)
  (Sorry. Could not find the manufacturer).

The following weather radio receivers have fewer features than emergency radios. Nonetheless, they are quite adequate to monitor weather warnings and alerts with S.A.M.E. technology:

- **Sangean** CL-100
  http://www.sangean.com/products/product_cate₅
  cid=12

- **Midland** WR-120 and WR-300
  https://midlandusa.com/product-
  category/weather-radios/emergency-
  preparedness/

- **AcuRite** 08525, 08535 and 08580
  http://www.acurite.com/emergency-weather-
  alert-noaa-radio

- **Oregon Scientific** WR608
  http://www.oregonscientificstore.com/Weather-
  Radios.buy

Manufacturers of Lightning Protection Products

- **Polyphaser** Lightning Protection Products
  http://www.polyphaser.com/

- **Alpha Delta** Surge Protectors
  http://www.alphadeltacom.com/pdf/prices_surge
  4.pdf

- **DX Engineering** Grounding and Lightning
  Protection Products
  http://www.dxengineering.com/search/departmei
  and-lightning-protection

- **Storm Copper Components Co.**
  http://www.stormcopper.com/stormbrochure/Ha:

- **Morgan Manufacturing**
  http://www.morganmfg.us/

Lightning Detectors

Lightning detectors are essential if you live in a lightning-prone area. They help forewarn you of impending lightning strikes.

You will find links to lightning detection manufacturers below. You should consult the first one in the list, before visiting the other sites, to acquaint yourself with the variety of lightning detectors on the market. This information will help you in making an informed choice.

Read this first: *Overview of Lightning Detection Equipment* by NLSI
http://www.lightningsafety.com/nlsi_lhm/detectors.html

- **Boltek** Lightning Detection Systems
  http://www.boltek.com/
  http://bolteklightning.com/

- **SkyScan** Lightning Detectors
  http://skyscanusa.com/
  http://www.skyscancanada.com/

- **StrikeAlert** Personal Lightning Detector
  http://www.strikealert.com/

- **Grainger** Lightning Detector
  bit.ly/1OTsdmo

- **Acurite**
  http://www.acurite.com/environments/lightning-detectors.html

- **Strike Guard**
  http://www.wxline.com/lightning_warning.php

- **Campbell Scientific**
  http://www.campbellsci.com/lightning

- **Aplicationes Technologicas** (Spain)
  Technologies for Preventive Lightning Protection
  http://lightningprotection-at3w.com/products/storm-detectors/s2xat5

- **WXLine** Wave Siren Station
  http://www.wxline.com/siren_station.php

- **Thor Guard**
  http://www.thorguard.com/products/

Tracking Lightning Online

It is essential to understand that there is a **delay** between the occurrence of lightning, and its display on the online sites. Also, geographical location of strikes on the maps is relatively *coarse*. Use them with caution, as lightning strikes may already be too close for your safety, even as you are watching the so-called "real time" displays on your screen. A quality lightning detector is strongly advised.

The following websites will help you to prepare early for lightning, much before your lightning detector begins to detect activity. The information provided by these sites will help you decide what actions to take to monitor lightning more closely.

- **VAISALA STRIKEnet** Lightning Verification Report (USA coverage) http://thunderstorm.vaisala.com/explorer.html

- **Environment Canada** (Canada coverage) http://weather.gc.ca/lightning/index_e.html

- **StrikeStar** (US coverage and near border regions of Canada) http://www.strikestarus.com/

- **NetWeather.tv** (UK coverage) http://www.netweather.tv/index.cgi? action=lightning

- **Network for Lightning and Thunderstorms in Real-Time** (World coverage) http://www.blitzortung.org/en/page_0/index.php

- **LightningRing** (World coverage) http://www.lightningring.com/map.html

References on Lightning Protection

Ham Radio Station Protection by PolyPhaser: a **must read**.

bit.ly/1ZprVW7

Lightning Protection by ARRL
http://www.arrl.org/lightning-protection

The Lightning Protection Institute
http://lightning.org/

Lightning Safety by the National Weather Service
http://www.lightningsafety.noaa.gov/safety.shtml

Station Ground by W8JI
This site contains a wealth of instructive diagrams and
pictures on lightning protection.
http://www.w8ji.com/station_ground.htm

CHAPTER TWO

The Antenna Tuner

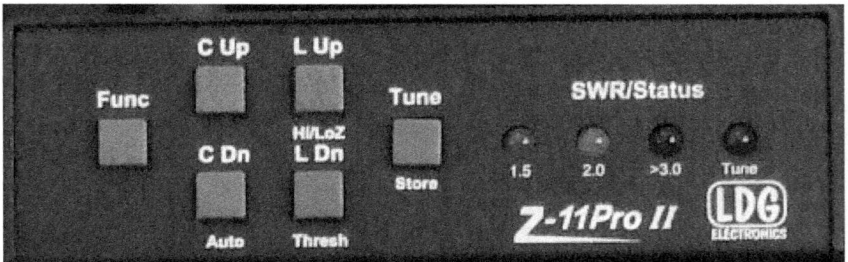

The antenna tuner, or "transmatch", is a variable impedance matching device. It is installed between the transmission line of a HF antenna system and a transceiver or transmitter. Its purpose is to make the antenna system "look like" a purely resistive load of 50 ohms in order to match the impedance of amateur radio transceivers at the antenna connector.

Basically, amateur radio operators use their tuners to vary inductance, along with input and/or output capacitance, in an effort to "tune out", or eliminate the capacitive and/or inductive reactance that may appear at the transmitter end of the transmission line.

Why match the antenna system impedance to the transmitter's output impedance?

Because maximum transmitted energy is transferred to the load (the antenna system) only when the load matches the characteristic impedance of the transmitter output, which is usually 50 Ohms.

Ideally, we want all the RF energy generated by the transmitter to reach the antenna. If the antenna system reactance is not prevented from appearing at the transmitter (by the tuner), we risk damaging the components of the final amplifying stage, due to high SWR.

Furthermore, if an "L-network" or "Pi-network" is used, every microvolt of RF energy received by the antenna system, which succeeds in making it down the transmission line to your tuner, will be transferred almost intact to your receiver circuit.

Please note that the tuner, when installed between the transmitter output and the transmission line...

- Will not eliminate the standing waves that may appear on the transmission line.
- Will not eliminate the portion of the signal which is

lost as heat within the transmission line due to the standing waves.

- Will not tune the antenna to resonance at the other end of the transmission line (an unfortunate popular misconception).

T-Network Tuners

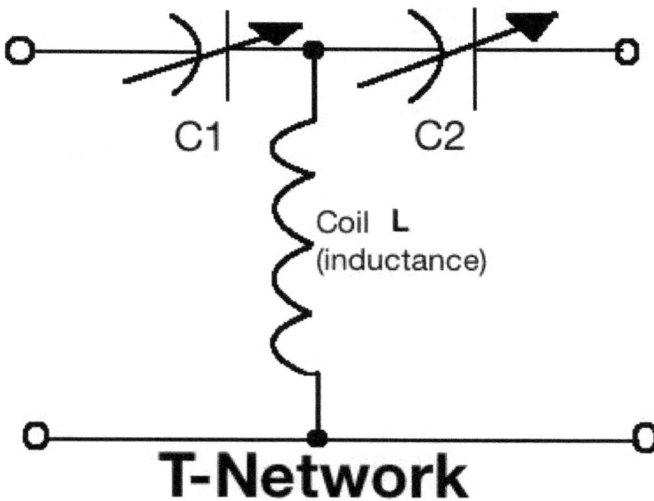

T-Network

Be aware that "T-network" antenna system tuners are the **least efficient** of all possible network configurations. Most inexpensive antenna tuners available commercially today are "T-networks". The MFJ-941D pictured here is an example.

It is true that the T-network configuration will usually do the job of transforming complex antenna system impedance, appearing at its output (antenna system side), to a 50 ohm impedance at its input (transmitter side). However, in the process of cancelling out unwanted antenna system reactance, the T-network will waste a prohibitive portion of the RF energy as heat, drastically reducing the efficiency of impedance transformation system. (ref: ARRL Antenna Handbook, chap.25).

Therefore, why continue to produce inefficient tuners?

Essentially, because they cost very little to manufacture. Consequently, they are relatively inexpensive, which makes them very attractive to unwary amateur radio operators.

The RF lost in T-networks when transmitting may not be sufficient to prevent your signal from being *heard*, especially when propagation conditions are favourable. However, in receive mode, the loss of RF energy is often catastrophic. Other amateurs will hear you calling, but you will often be unable to hear them respond to your call.

L-Network Tuner

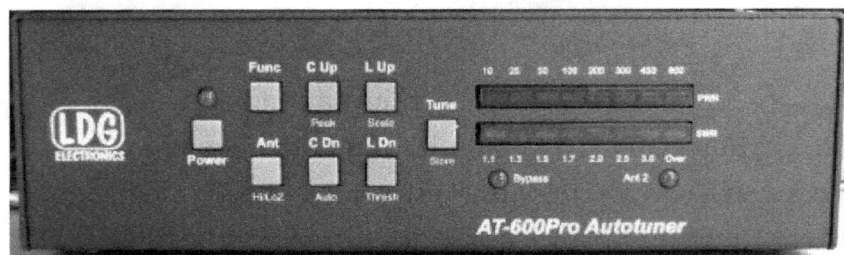

L-networks require very large capacitance values, in the thousands of picofarads (pf) at low frequencies. Variable capacitors with this much capacitance are extremely expensive.

The LDG AT-600Pro Autotuner effectively solves the problem of cost, by taking full advantage of the flexibility, efficiency and effectiveness of the L-network configuration. It automatically selects the proper L-network configuration to adapt itself to high or low impedance situations, and switches fixed values of capacitance, in or out, to fine-tune for a valid match.

As its name suggests, it will handle up to 600 watts of power on SSB and CW, and up to 250 watts on 6 meters. It interfaces directly with my IC-7200 transceiver which supplies the 12VDC it needs to run. Under this setup, the tuner operates completely automatically.

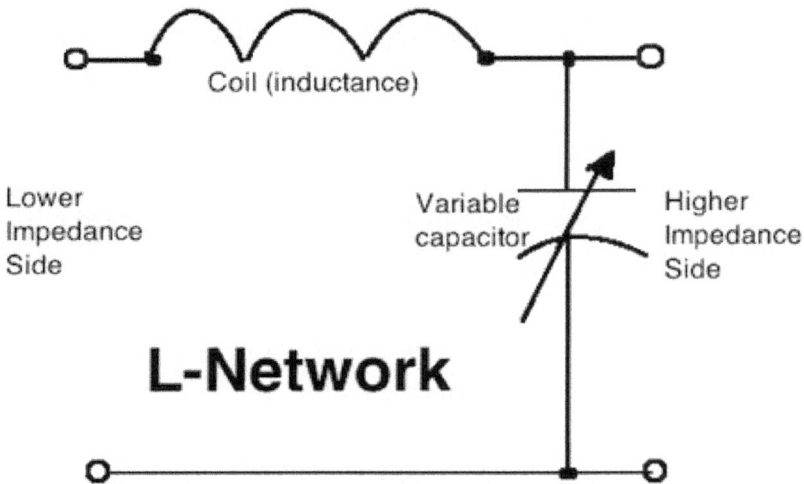

Coil (inductance)

Lower
Impedance
Side

Variable
capacitor

Higher
Impedance
Side

L-Network

The AT-600 switches on the configuration, illustrated above, to match the low impedance (50 ohms) of the transceiver output terminal to a higher antenna system impedance detected by the tuner. On the other hand, when the tuner detects that the antenna system impedance is lower than 50 ohms, it *reverses* the configuration so that the antenna is connected, instead, to the lower impedance side of the "L-network".

That is very clever, and the reason I decided to replace my MFJ941D with the AT-600Pro tuner for multiband operation with my Carolina Windom antenna.

I also have a LDG Z-11Pro II that I use when I travel. Its small footprint and light weight make it a great companion to my Flex-1500. The Z-11Pro II is just as efficient as its big brother the AT-600 Pro. It can handle power levels between 0.1 and 125 watts SSB and CW, and 30 watts on PSK and other digital modes. It will even handle 100 watts on 6 meters comfortably.

The LDG antenna tuners, mentioned above, are made for coaxial transmission lines. There are very good reasons for this. If you use open wire transmission lines, I highly recommend that you read DJ0IP's excellent article *Symmetrical Tuners for Open Transmission Lines* (http://www.dj0ip.de/antenna-matchboxes/symmetrical-matchboxes/). Read it to the end. You will find it to be an invaluable reference work.

Antenna Tuner Manufacturers

I include below a partial list of antenna tuner manufacturers that I recommend above others:

- **LDG** Tuners

(http://www.ldgelectronics.com/c/252/products)

- LDG Product Manuals & Specs
(http://www.ldgelectronics.com/c/261/product-manuals).

- **SGC** Smartuner™
(http://sgcworld.com/store/page2.html).

- **Palstar**
(http://www.palstar.com/en/at-500specifications/).

- **TEN-TEC** Tuners
(http://www.rkrdesignsllc.com/products/tuners-1/).

QRP Tuners

- **EMTECH** ZM-2 ATU
(http://emtech.steadynet.com/zm2.shtml).

- End-Fed Half-Wave Tuner **Kit**
(http://www.qsl.net/w6dps/EFHW.html).

Finally, there are the ubiquitous MFJ Tuners. I do not recommend them, as they are very inefficient "T-network" configurations. As I explained above, their deceptively low price makes them appear attractive, but they simply do not measure up to expectations. Unfortunately, most MFJ tuner

owners will take years to notice the shortfall in performance, if ever.
(http://www.mfjenterprises.com/Categories.php?sub=0&ref=5).

CHAPTER THREE

The SWR/PWR Meter

The SWR/PWR meter is a popular antenna accessory. There are many brands of HF SWR/PWR meters on the market such as Ameritron, Comet, Daiwa, Diamond, MFJ and Jetstream, to name a few. They have one characteristic in common: they are factory-assembled and, as such, can be considered as "plug & play" consumer products.

But such a wide variety of available products can be overwhelming.

I believe that the best way to ever become able to make an informed choice is to build one. By doing so, you get acquainted with the problems involved and the inevitable compromises you (or the manufacturer) must make to keep the cost down while achieving an "acceptable" level of performance. But what's "acceptable"? I would suggest:

1. Whatever exceeds the performance of the meter you have built and used on the air for a while.

2. Whatever you find more convenient to use than the SWR/PWR meters already incorporated in most modern transceivers.

Therefore, let's begin by looking at a simple, economical, yet reasonably accurate "outboard" SWR/PWR meter solution for low power applications.

A Simple Homemade Meter

My preference goes to the *"Resistive SWR Bridge"* by G3ROO & G4WIF because it has a unique feature: its design will protect the final amplifier transistor from high SWR situations, even in the presence of a dead short or open circuit.

This meter is ideally suited for QRP operations. The Web page contains a circuit diagram and a parts list, as well as a brief and easy to understand explanation on the theory of operation. (http://www.interalia.plus.com/q_tech15.htm). Note: the author indicates that the information will eventually be moved to (http://www.gqrp.com/technical15.htm).

Meter Kits

Now that you know you can get relatively accurate readings from a homemade meter, let's have a look at more sophisticated and accurate solutions in kit form.

- **YouKits** DP-1 QRP Digital Power & SWR Meter This meter is also designed for QRP operations. Although it is made by YouKits, this instrument is only sold factory assembled and ready to use. The DP-1 will measure power from 5 mW to 25 W for frequencies between 1 and 60 MHz. Its accuracy is +/- 5%. (http://youkits.com/products/youkits-dp-1-qrp-digital-power-swr-meter).

- **Elecraft**-W1 140W Computing Wattmeter and SWR

Bridge

The Elecraft W1 is a versatile microprocessor-based RF power and SWR meter which can be used with any transmitter ranging from QRP levels to 140 watts output, and from 1.8 to 30 MHz. PC software is supplied if you prefer a graphical user interface instead of the built-in bar-graph LEDs. (http://www.elecraft.com/mini_module_kits/mini

- **Fox Delta** Meters

  Fox Delta offers numerous models of SWR/PWR meters. I especially like the "SWM3-0915 Dual Channel LCD SWR Meter" (kit) by I2TZK, K7SFN & VU2FD. It uses a PIC16F877A micro controller for accuracy. Boot-loaded firmware and PC software are provided free of charge. The latter generates an exceptionally good-looking graphical user interface (GUI). This kit is intended for experienced kit builders but it is assembled, free of charge, for senior radio operators. A very nice touch. (http://foxdelta.com/products/swm3.htm).

- **Oak Hills Research** WM-2QRP PWR/SWR Meter Kit

  The Oak Hills Research WM-2 was designed specifically for the QRP operator. The unit operates from 300 KHz to 54 MHz. It will measure forward and reflected power at QRP levels down to 5 mW.

(http://ohr.com/wattmeter.htm).

Factory-assembled SWR/PWR Meters

For your convenience, here is a list, in alphabetical order, of manufacturers producing SWR/PWR meters suitable for use during amateur radio operations.

- **Ameritron** SWR/Wattmeters
  (http://www.ameritron.com/Categories.php?
  sub=0&ref=32).

- **Comet** CMX-2300
  (http://www.cometantenna.com/amateur-
  radio/swr-meters-analyzers/comet-meter/).

- **Daiwa** CN-101 Economy Series HF/VHF Bench
  Meter
  (http://www.cometantenna.com/amateur-
  radio/swr-meters-analyzers/daiwa-meters/)

- **Diamond Antenna** Power/SWR Meters
  (http://www.diamondantenna.net/Product_Catalo

- **Jetstream** JTWHF (and many more models)
  (http://www.jetstream-usa.biz/index.php?
  cPath=8100)

- **MaCo Antennas** Maco 2400
  (http://www.macoantennas.net/accessorypages/M

- **MFJ Enterprises** MFJ-812B (and many more models)
  (http://www.mfjenterprises.com/Product.php?productid=MFJ-812B).

- **Palstar** PM2000A
  (http://www.palstar.com/en/pm2000a/).

- **Telepost** LP-100A Digital Vector Wattmeter
  (http://www.telepostinc.com/lp100.html).

- **Vectronics** PM-30
  (http://www.vectronics.com/Product.php?productid=PM-30).

Laboratory Grade Instruments

If prefer high precision lab instruments, and do not mind paying a usually hefty price, here is an example of what you might find acceptable, short of buying a Bird wattmeter. It is designed to meet the expectations of the most demanding amateur radio operator.

- **Alpha** 4500A Series Wattmeters
  The Alpha 4510A Series High Frequency RF Power

Meter (a picture of which is at the beginning of this chapter) is a **laboratory grade** instrument capable of measuring and displaying transmit forward power, reflected power, delivered power, and SWR at continuously variable power levels between 0 and 3 kilowatts (0 to 5 KW for the 4520A model). Both models cover amateur radio frequency bands between 1.8 MHz and 54 MHz.
(http://www.rkrdesignsllc.com/-8/).

CHAPTER FOUR

Antenna Analyzers

If you are as passionate about experimenting with your antenna as I am, I strongly recommend that you acquire an antenna analyzer. If you do not have the money to buy one, try to borrow one, or build the inexpensive kit described below.

What advantages does an antenna analyzer provide? It…

- removes the guesswork from antenna experimentation;

- dramatically reduces the time you spend on an antenna project before meaningful results are reached;

- increases your rate of success considerably;

- makes antenna experimentation a rewarding activity in terms of self-satisfaction;

- enables you to match an antenna system to your needs and preferences, rather than adopting the "one size fits all" approach of most commercial antennas;

- encourages experimentation and innovation.

RigExpert AA-230 PRO

I have been using this instrument on all my antenna experiments since 2012. It is very accurate. Its RF generator is very stable and is not influenced by even the most complex antenna impedances. Consequently, the displayed results are also very stable. It has a variety of functions to test antennas and coaxial feedlines for frequencies ranging from 0.3 to 230 MHz.

It also has a unique feature that I like very much. It's called SWR2Air™. This feature enables me to make adjustments to the antenna outdoors, without requiring the help of someone in the radio room indoors to report the SWR fluctuations as I make adjustments. Here is how it works. The analyzer is connected to the transmission line, in the shack, where it is put in SWR2Air™ mode and set to transmit SWR information on a frequency of my choosing. I can then go outside and make adjustments to the antenna. I bring my handheld FM XCVR with me to receive SWR information signals generated by the AA-230 PRO. The analyzer transmits the SWR info in the form of beeps, which get shorter as SWR falls, and lengthen as SWR rises, thus guiding me in my adjustments. This feature alone is priceless. I cannot imagine experimenting with antennas without it.

The LCD display is sharp and easy to read, even when displaying graphs. In addition, the AA-230 PRO comes with software called AntScope which enables you to visualize the SWR curve plotted on a large graph displayed on your PC's larger screen. The software allows you to keep a record of the graphs, which serve to document the results of your experiments, as you progress in your antenna project, as well as for future reference.

I briefly demonstrate, in this video (see link), how I used the AA-230 PRO's to help build a homemade 6-meter folded dipole: https://youtu.be/qbje13U6hC4

Talking to the AA-230 PRO

For the more advanced users, RigExpert lets you "talk" to the AA-230 PRO by connecting it to your PC. This useful and innovative feature is nothing less than outstanding. The full set of commands is given on this page of the manufacturer's website:
http://www.rigexpert.com/index?f=aa_commands

How Does the AA-230 PRO Work?

This short article will give you an inside look at the technology behind the engineering of the AA-230 PRO.
http://www.rigexpert.com/index?s=aa230&f=inside

For a complete description of all its features and capabilities, I suggest you download and read its manual (PDF):
http://www.rigexpert.com/files/manuals/aa230230pro52(

Read This Before Buying An Analyzer

The RigExpert website offers two interesting articles. Together, they will help you understand what to expect from antenna analyzers now available on the market.

I strongly recommend that you read "*A short review of antenna and network analyzers*" BEFORE you buy an antenna analyzer. It's an "eye opener". I could not have written a more revealing description of the analyzer situation myself. This article is indispensable information that will make you an informed (forewarned) buyer. You will know what to expect and what you are paying for, regardless of the brand and model you ultimately choose to buy. http://www.rigexpert.com/index?s=articles&f=aas

Next, you should read, "*A very cheap antenna analyzer for HF bands*": http://www.rigexpert.com/index?s=articles&f=cheapaa

Note that the AA-230 PRO was recently replaced by a new model, the *RigExpert AA-230 ZOOM Antenna and Cable Analyzer*, which covers 100 kHz to 230 MHz. http://www.rigexpert.com/index?s=aa230zoom

My Previous Analyzer

I had bought a ZM-30 from Palstar in September 2010. It worked perfectly for a few months, and then began giving faulty impedance readings. I wrote to 'Support' describing the situation. They responded that I should simply restore the unit to its default values stored in its EEPROM, and then run the unit through the setup and calibration procedures given in the manual. It solved the issue only temporarily, as the problem returned within a few months. As it would still display correct SWR readings, I decided to NOT return it to Palstar for expensive repairs. I had read the negative reviews of other hams whose units were never properly fixed, even after spending a great deal.

Palstar has given up. The ZM-30 is no longer in production.

Analyzer Kit

For those of you who are not yet ready to spend hundreds of dollars on an antenna analyzer, here is an interesting DIY analyzer idea that evolved from a project by K6BEZ.

This is a very capable analyzer that one can build for less than USD$50. It covers HF frequencies between 1 and 30 MHz. This site will help you learn how the project got started, its success at Pacificon 2013, and will walk you through the various developmental stages. Most importantly, however, K6BEZ provides a complete list of all the parts, as well as a detailed assembly guide (PDF). https://sites.google.com/site/k6bezprojects/antenna-analyser

A Book on Antenna Analyzers

If you would like to read more on antenna analyzers and their many uses, I recommend the ARRL book, Understanding Your Antenna Analyzer, by W1ZR, Joel R. Hallas.

Antenna Analyzer Manufacturers

Here is a partial list of manufacturers of antenna analyzers suitable for HF antenna systems. Before considering any of the following products, please read *"A short review of antenna and network analyzers"* mentioned previously in this chapter.

- **RigExpert** Antenna Analyzers
  http://www.rigexpert.com/

- **AEA Technology** 140–525 Analyzer

http://www.aeatechnology.com/swr-meters

- **Array Solutions** AIM4300 by W5BIG
  http://w5big.com/AIM4300.htm

- **Bird Technologies** SiteHawk SK-200-TC
  http://birdrf.com/Products/Analyzers/Site-Hawk.aspx

- **Comet** Antenna Analyzer CAA-500MARKII
  http://bit.ly/1WUWeUQ

- **iPortable** iP30z Analyzer
  http://iportableus.com/_mgxroot/page_10743.htm

- **MetroVNA** Pro Touch
  http://www.metrovna.com/description/

- **MFJ Enterprises** MFJ-225 (least recommended)
  http://www.mfjenterprises.com/Product.php?productid=MFJ-225

CHAPTER FIVE

Amateur Radio Tower Guide

The selection and installation of an amateur radio tower must be planned carefully. When you do your research properly, as explained below, your tower will give you many decades of safe and reliable service.

My 15 meter (48 feet) free-standing, heavy-duty tower has been serving me loyally for almost 40 years. Installed first in 1977, it was dismantled and reassembled in 1987, when we moved to the country, from the suburbs of Montréal (QC, Canada), and continues to serve my needs. I mainly use it as one of the supports for my Carolina Windom 80 and for my 160-meter inverted-L. My 2-meter vertical antenna sits on top of the tower.

Essential Considerations

You must first ensure that your local town authorities allow ham radio towers in the neighbourhood where you live. If they are permitted, you are one of the lucky ones.

Armed with a copy of the local bylaws and ordinances, you can then proceed to select the one which conforms to local regulations, and still meets all of your own needs. Among the first essential criteria to consider:

- Your tower must not be installed where it could fall on power lines.

- You must take steps to prevent children from climbing the tower (i.e. a locked fence around the base).

- You should not install a tower where it could fall on the neighbor's property.

Other Practical Considerations

The tower should be sturdy enough to withstand the wind load that the size and weight of the antenna(s) will impose on the tower. Tower manufacturers specify this characteristic in the tower specifications. Therefore, you must take into consideration:

- Local climate conditions likely to be experienced. Key factors are maximum winds recorded over the past 30 years for your area, and the likelihood of freezing rain episodes.

- Antenna wind load of present and future antennas. This information is available from the antenna manufacturers.

Plan for the future carefully. You may only be able to afford a small HF yagi tri-bander to start, but you will eventually want to add a multi-element VHF yagi, a UHF yagi and/or a larger multi-element HF yagi in the years ahead.

You can either decide on a heavy-duty tower to start, or plan to replace your first light duty tower in a few years. There is simply no doubt that you will want to add more antennas to your tower as you become more experienced — that is a given.

If this is your first tower, you should ask for the opinion of other hams in your neighborhood, and consult the members of your local club. They will supply priceless advice, and probably even offer to help you install your tower.

Tower Types

There are four main types of towers. Each of them has its specific installation requirements. They are generally available in light, medium or heavy-duty materials.

- **Guyed.** In terms of "cost per foot", the guyed tower is the most economical. However, you must have enough space for the guy wires. Guy anchors should be installed away from the base of the tower at a distance varying between 60% to 80% of the tower height, depending upon the manufacturer's instructions.

- **Free-standing.** This option is only moderately expensive, and has a minimal ("lean") impact on the visual environment.

- **Fold-over or tilting.** These towers have a convenient hinged base and/or hinged sections. They are more expensive, but very practical if you want to be able to work on your antennas without having to climb the tower each time.

- **Crank-up.** These can be hand cranked or motor driven, and are certainly the most expensive. They are also the most practical.

New towers generally include detailed installation

instructions provided by the manufacturer. They must be followed meticulously. Remember to take into account any additional specifications that your town's technical department might have imposed.

Final Recommendation

Before buying a *used* amateur radio tower, inspect it for rust (especially paint covered rust) and wind damage (warping). This is the time when you should reach out to a few experienced operators to help you with the selection and installation.

Tower Manufacturers

The following list of amateur radio tower manufacturers does not pretend to be complete. If you are planning to buy a tower, do take the time to visit each site. Examine carefully what they have to offer, and pick the make and model that will best fit the present and future requirements you identified by following the recommendations outlined above.

- **HEIGHTS TOWER SYSTEMS** (www.heightstowers.com).

- **GLEN MARTIN Engineering Inc.** (glenmartin.com/products-2/amateur-radio).

- **ROHN Products International LLC** (www.rohnnet.com/).

- **ALUMA Tower Company Inc.** (www.alumatower.com/index.html).

- **US Tower Corp.** (www.ustower.com/).

- **TITANEX GmbH** (www.titanex.de/frames/contact.html).

- **ComTrain LLC—Towers and Monopoles** (www.comtrainusa.com).

CHAPTER SIX

The Low-Pass Filter

The low-pass filter (LPF), shown above, is the one I used before television stations began broadcasting digital signals instead of analog. It is an adjustable LPF that was made by Taylor Communications in Canada (circa 1975), which no longer exists. None of the LPFs on the market today are adjustable.

LPFs are not really necessary anymore in North America, except in rare circumstances, such as when transmitting at high power with neighbors nearby.

So, why mention them at all?

There are still some regions of the world where television

signals are still analog. In such cases, an LPF is necessary when transmitting on HF, especially if your neighbors are likely to watch channels 2 and 3 using an antenna.

Before I go on, please note that *most interference, even to televisions, is not caused by spurious emissions from a transmitter, but is instead caused by overload of the affected equipment by the fundamental-frequency signal from the transmitter.*" (Source: Ed Hare, W1RFI, ARRL Lab http://www.arrl.org/forum/topics/view/220)

The LPF's Role

In the 'olden days', low-pass filters were used to reduce/eliminate interference to the reception of TV channels 2 and 3. Such interference is called TVI. It occurs when the second harmonics of the 10-meter band signals would fall well within the channel 2 allocated frequency band (54-60 MHz).

- The second harmonic of 28 MHz, at the lower end of the 10-meter band, is 56 MHz.

- The second harmonic of 29.7 MHz, at the upper end of the 10-meter band, is 59.4 MHz.

In addition, the 4th harmonic of 20-meter signals, and the

8th harmonic of 40-meter signals, would also fall within TV channel 2 allocated frequency band and cause interference, albeit to a somewhat lesser extent than the 10-meter second harmonics.

However, nowadays, North American TV transmissions are digital instead of analog. TV channels that used to be in the lower VHF spectrum have been moved up into high VHF and UHF. They are thus further removed from second and third harmonic emissions of HF transmitters. See W1RFI (ARRL Lab) comments at: http://www.arrl.org/forum/topics/view/220.

If you operate in a region of the world where TV receivers are still analog, and TV channels are still in the low VHF segment, then here are examples of low-pass filters still sold today. Be aware that a poorly designed LPF, made with cheap components, will not fulfill its intended purpose, which is to prevent TVI. Furthermore, a cheaply made LPF will introduce significant RF losses, which will be most noticeable on receive. You definitely do not want to deprive yourself of single microvolt of RF energy your antenna has managed to capture!

References On TVI and LPFs

Several years ago, HS0ZEE wrote an article on TVI low-

pass filters. I believe it gives a good explanation of how harmonic radiation from a HF transmitter can interfere with TV reception (before TV stations switched over to digital on higher frequencies).
http://www.hs0zee.com/Amateur/TVI/TVi%20Filters.htr

You may also wish to look at W3NQN's filter for QRP. His document will help you understand how a low-pass filter is built.
http://www.gqrp.com/Datasheet_W3NQN.pdf.

Last, but not least, please keep in mind that an LPF **must** be used where it is **always** presented with a pure resistive **50 ohm** impedance on its antenna side. In most cases, this means that the LPF must be installed between your transceiver and the antenna tuner. If the impedance present on the antenna side of the LPF contains reactive components (inductive or capacitive), the LPF will **not** perform as expected. In some mismatched situations, the LPF will begin to introduce RF losses and may even suffer permanent damage by RF arcing between components inside.

Commercially-Built LPFs

Not all low-pass filters are created equal.

- **DX Engineering** Bencher Low Pass Filters YA-1 (http://www.dxengineering.com/parts/bnr-ya-1).

- **PALSTAR** FL-30 (http://www.palstar.com/en/filters/).

- **MFJ** (least recommended)
  http://www.mfjenterprises.com/Product.php?productid=MFJ-702B (LOW PASS FILTER, 1-30 MHZ, 200 W)
  http://www.mfjenterprises.com/Product.php?productid=MFJ-704 (LOW PASS FILTER, 1.5 KW).

High-Pass Filters (HPF)

The HPF is designed to reject (attenuate considerably) RF signals below 56 MHz received by a TV antenna. When television reception is interfered with (image and/or sound quality degraded intermittently), an HPF should be inserted between the TV antenna cable and the television receiver. An HPF will usually only be required if the television owner, in your neighbourhood, is experiencing TVI from your transmissions. The demand for HPFs has dropped considerably in recent years. They are becoming harder to find as many manufacturers, such as Ameco and Drake, have stopped producing them.

HPF Manufacturers

- **Vectronics** HPF-2 (US)
  (http://www.vectronics.com/Product.php?
  productid=HPF-2).

- **MFJ** HI-PASS Filter (US)
  (http://www.mfjenterprises.com/Product.php?
  productid=MFJ-711B).

- **SPECTRUM Communications** (UK)
  (http://www.spectrumcomms.co.uk/TVI_Filters.ht

- **QUASAR Electronics** (UK) 1145KT TV High-Pass
  Filter (KIT)
  Requires assembly and soldering.
  (https://www.quasarelectronics.co.uk/Item/smart-
  kit-1145-tvi-high-pass-filter).

A Homemade HPF Solution

If needed, here are two links to instructions on how to
build an HPF.

- http://homepage.ntlworld.com/rg4wpw/filters.ht
- http://www.uksmg.org/content/filter.htm

Reference Books On Radio Frequency Interference (RFI)

- The ARRL RFI Book, 3rd Edition (http://www.arrl.org/shop/The-ARRL-RFI-Book-3rd-Edition/).

- The RFI Pocket Guide (http://www.arrl.org/shop/RFI-Pocket-Guide/).

CHAPTER SEVEN

HF Radio Wave Propagation

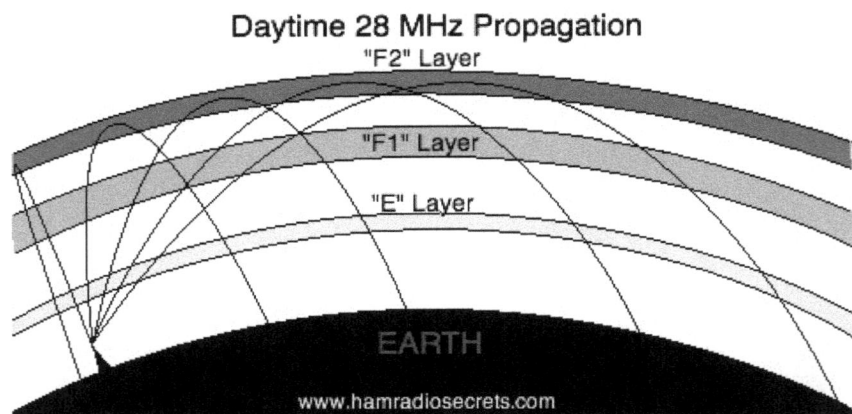

Daytime 28 MHz Propagation

About every 11 years, the sun enters a period of increased activity. Large nuclear explosions begin to occur in its upper atmosphere. These are commonly called "sunspots". The "solar wind", generated by the sun's explosions, impacts the earth's upper atmosphere, forming ionized layers in a region of the upper atmosphere called the ionosphere.

At night, the "F2" layer (about 190 miles/300 km above the earth) acts like a mirror to HF radio signals.

During the day, the "F1" layer (about 120 miles/200 km above the earth) takes over the role of bending HF signals back toward the earth's surface, but for HF frequencies mostly above 6 MHz. Therefore, the "F1" layer is the active daytime deflector of amateur radio HF signals from 40 meters up to 10 meters.

These two ionized layers play an essential role in enabling amateur radio stations to communicate over hundreds of miles on the lower amateur bands, and thousands of miles on the higher frequency bands, sometimes permitting contacts between opposite sides of the planet.

Sunspot Cycle 24

Sunspot Cycle 24 began slowly subsiding in 2014. Even so, it will still produce some good DX opportunities for a few years. In June 2009, an international panel of experts had predicted that Solar Cycle 24 would likely not produce a smoothed sunspot number (SSN) exceeding 90.

In fact, it laboriously reached a peak of 81.9 in April 2014.

In contrast, Cycle 23 had produced a maximum SSN

around 160 in 2001 which, in itself, was less than the maximum observed during each of the two previous cycles.

In other words, we were told not to expect too much out of solar cycle 24. The experts were right.

Nevertheless, cycle-24 provided amateur radio operators all over the world with a number of good to excellent signal propagation conditions at times. Now that it is over the top, we have to watch the sun's activity more closely to make the most of it.

It is so quiet on HF during a solar sunspot minimum that *any* sunspot activity, and its beneficial effect on the formation of the "F" layers, is very welcome indeed.

Monitoring the Sun's Activity

Paul, N0NBH's website is one of the most informative sites on solar radiation and the sun's influence on signal propagation. If you are at all interested in acquiring a better understanding of the sun's impact on our hobby, this is one site I recommend without hesitation. http://www.hamqsl.com/solar.html.

The smoothed sunspot number (SSN) represents the

average number of sunspots observed over a whole year. Generally speaking, the higher the number of sunspots, reported as smoothed sunspot number (SSN), the higher the *Maximum Usable Frequency* (MUF).

SSN, A index, K index, Solar Flux... What do the numbers mean? If you are interested, I recommend that you read this article by K9LA: http://www.arrl.org/the-sun-the-earth-the-ionosphere.

For regular information on solar activity and its effect on propagation, I recommend the ARRL Propagation Forecast Bulletin produced by Tad Cook, K7RA, which you will find at: http://www.arrl.org/w1aw-bulletins-archive-propagation.

I also recommend a visit to Kevin's website (VE3EN): http://solarcycle24.com. It offers an impressive amount of information on solar activity. It is a true feast for the eyes and the mind.

Amateur Radio Beacons

Learning how to monitor amateur radio beacons is the secret to successful and plentiful DX contacts. Beacons provide early and measurable indications of propagation conditions. By being diligent, and patient, you can send out a CQ as soon as you begin to hear a beacon. In doing so, you will often be the one to start a pileup, instead of trying to fight your way in.

Amateur Radio Beacon Frequencies

The *Northern California DX Foundation* (NCDXF) operates eighteen amateur radio beacons on five continents which transmit in successive one-minute intervals on 14.100, 18.110, 21.150, 24.930 and 28.200 MHz.

The NCDXF beacon call sign and the first dash is sent at 100 Watts. The remaining dashes are sent at 10 Watts, 1 Watt and 0.1 Watts. These two latter power levels are very useful for QRPers.

Almost all ten-meter beacons transmit between 28.190 MHz and 28.300 MHz. You will find a comprehensive list at: http://userpages.troycable.net/~wj5o/bcn.htm.

Six-meter beacons are found mostly between 50.0 MHz and 50.1 MHz, with a concentration between 50.06 MHz and 50.08 MHz. One example is W4CLM/B transmitting 30 Watts continuously into a vertical on 50.065 MHz (+/-) from location EM74 (Atlanta, GA.).

Beacon Monitoring Antennas

Ideally, you should use the same antenna that you normally use for HF communications. If you use a different antenna, it will likely not have the same pattern of radio wave capture. It may give you information on paths of propagation that you may not be able to cover with the HF antenna you normally use to make contacts with.

I use my homemade 160-meter inverted "L" as a general-purpose multiband antenna. If you use a directional antenna to monitor the beacons (i.e. multi-element beam), you will only effectively hear the beacons that lie in the direction your beam is aimed at. On the other hand, that may be just be what you want.

Beacon Monitoring Software

Monitoring amateur radio beacons manually can quickly tiresome. But with today's technology, it is unnecessary. Many software programs have been written for a wide range of personal computer operating systems. Using software to control your receiver automates the beacon scanning process and frees you to devote your time to other more pressing activities.

The *Northern California DX Foundation* lists a number of programs that have been tested, and proven to work as advertised. You will find them listed under *"Tools for Listeners"* at:
http://www.ncdxf.org/Beacon/BeaconPrograms.html.

Other Means of DX Monitoring

Many amateurs monitor their local packet DX Cluster node for DX activity, allowing them to decide whether, or not, to join in the action. Others log on to DX nets, seeking a chance to make brief contact with participating DX stations.

Obviously, monitoring beacons to alert you of band openings will let you become the "action", which will later be reported on DX cluster nodes.

Non-Amateur Radio Beacons

The time and frequency standard stations (listed below) will give you some indication of propagation conditions at their operating frequencies. But, keep in mind that they transmit at much higher power levels than the amateur radio maximum legal power limit. In other words, when you can receive a signal from CHU or WWV, it does not necessarily mean that the nearby amateur radio band is usable by most amateur radio operators.

The CHU Canada time signal transmitting station is located 15 km southwest of Ottawa, Ontario, Canada at 45° 17' 47" N, 75° 45' 22" W. Main transmitter powers are 3 kW at 3330 and 14670 kHz, and 10 kW at 7850 kHz. Individual vertical antennas are used for each frequency.

WWV transmits from Fort Collins, Colorado, and WWVH, from Kauai, Hawaii, on 2.5, 5.0, 10, 15 and 20 MHz. To supplement your monitoring of amateur radio beacons, you can listen to WWV broadcast the latest solar-flux index at 18 minutes past the hour, and at 45 minutes past the hour on WWVH. Again, generally speaking, the higher the solar-flux index, the higher the MUF.

Hf Propagation Monitoring Software

I prefer VOACAP (Voice of America Coverage Analysis Program), which is a free professional HF propagation prediction *online* software from NTIA/ITS. In my opinion, it stands as the uncontested best.

Read the VOACAP Quick Guide (free) first, to get the most out of this exceptional "HF Propagation Prediction and Ionospheric Communications Analysis" service. http://www.voacap.com/

- VOACAP Online (free)
  http://www.voacap.com/prediction.html

- VOACAP Propagation Planner (free)
  http://www.voacap.com/planner.html

- The 24/7 real-time HF propagation monitoring cluster, which covers the entire HF spectrum from 1.8 MHz to 28 MHz, is based on Skimmer and

Skimmer Server.
http://www.voacap.com/skimmer/index.php

- VOACAP Primer (free)
  http://hamwaves.com/voacap.primer/en/index.ht

PC Stand-Alone
Propagation Monitoring Software

- **DX Toolbox** by Black Cat Systems
  Quoting from their website: *"DX Toolbox searches the web for you, gathering information on solar and geomagnetic conditions that affect radio propagation. It also features several propagation forecasting tools, allowing you to quickly and easily estimate current HF (Shortwave) propagation conditions between any two locations in the world"*.
  http://bit.ly/1P5QPoD

- **DXLab** (freeware)

- Quoting from their website: *"DXLab is a freeware suite of eight interoperating applications that can be installed independently in any order. When multiple applications are running, they sense each other's presence and automatically interoperate to support your Amateur Radio DXing activities"*.

http://www.dxlabsuite.com/

- **Ham CAP 1.9** by Afreet Software (freeware)
  http://www.dxatlas.com/HamCap/

- **Strategic Communications—ACE-HF PRO** by
  LONG WAVE inc.
  http://www.longwaveinc.com/ace/ace-hf-pro/

- DXZone Propagation Prediction Software for Ham
  Radio is a curation of 41 useful links.
  http://www.dxzone.com/catalog/Software/Propa

More About The Author

Claude Jollet is a self-proclaimed *incorrigible eclectic*. His day job was in operational meteorology for 31 years. He also has a degree in Management Information Systems and Computer Programming.

He had been sharing his passion, knowledge and know-how on websites since 2005, and decided in 2015 to become a self-publisher of e-books.

His website focusing on **amateur radio** is:
HamRadioSecrets.com.

His Other Websites

He is also the author and webmaster of the following websites:

www.Weather-In-Canada-Observer.com

www.Meteo-NDP.info (French)

www.Claude-Jollet-Ecrivain.info (French)

www.ClaudeJollet.com where he "bloomed out of anonymity" (his words).

You can follow Claude on Twitter, get in touch on Facebook, connect with him on LinkedIn, or e-mail him at: ve2dpe@hamradiosecrets.com.

Other Books In This Series

This e-book is a compendium of a four-book series on amateur radio HF antennas. They are individually available here in all popular formats: PDF, Kindle, ePub and in Apple's iBooks Store. The four other titles in the series are:

1. Book One: *Introduction* [to Amateur Radio HF Antennas]
2. Book Two: *HF Antennas For Limited Space*
3. Book Three: *Homemade HF Antennas*
4. Book Four: *HF Antenna Accessories*

All the e-books in this series are deliberately short and to the point. They are devoid of unnecessary prose that would only distract and prevent the reader from acquiring practical and immediately useful knowledge.

If you would like to be automatically informed when the next e-books in the series and their updates are released, visit his Ham Radio Blog and subscribe to its RSS feed.

May We Ask?

Have you found this e-book helpful? Please spread the word on your favorite social media channel. Have you found it to be of no particular use to you? Spread the word anyway! Others may not possess your level of knowledge and know-how. Let them know of this e-book. They might thank you for the tip.

Dedication

This e-book is dedicated to all amateur radio operators who love and long to share their knowledge and know-how about the *King of Hobbies*.

Acknowledgments

My many websites testify to my love of writing about my various passions, and areas of expertise. But my decision to publish e-books — and eventually their printed versions — comes from my exposure to the exceptional community of fellow website owners on the SBI! Forums. Among them, Harvey Chapman has influenced me the most. I thank him for his continuous contributions of priceless information which gave me the tools to develop the patience needed to undertake the painstaking work required to become a successful self-published author.

I learned how to promote my services and my writings, online and successfully, from a man named Ken Evoy. I will be forever in his debt.

I have to tip my hat to the people at Druide Informatique for their exceptional *Antidote 9* corrector. This software forces me to think how best to say what I want to communicate by making me go through a rigorous multi-step self-editing process with each of my texts. The result is *near perfect* and saves me hundreds of dollars in professional editing services. I may eventually hire a professional editor anyway — after my self-edited e-books have generated enough income — because, honestly, it requires a great deal of tedious work!

Finally, I need to thank my wife for humoring me while I toil behind closed doors on these e-books, with strict orders not to be disturbed, barring a fire or an earthquake. It works … often enough.

CONTENTS